Microwave Sensors for the Dielectric Characterization of Biological Cells

Vom Promotionsausschuss der
Technischen Universität Hamburg
zur Erlangung des akademischen Grades
Doktor-Ingenieur (Dr.-Ing.)
genehmigte Dissertation

von

ALEKSANDAR SAVIĆ

aus

BANJA LUKA

2023

1. Gutachter: Prof. Dr.-Ing. Arne F. Jacob

2. Gutachter: Prof. Dr.-Ing. Christian Damm

Prüfungsvorsitz: Prof. Dr.-Ing. Hoc Khiem Trieu

Tag der mündlichen Prüfung: 20. April 2023

Bibliografische Information der *Deutschen Nationalbibliothek*
Die Deutsche Nationalbibliothek verzeichnet diese Publikation in der Deutschen Nationalbibliografie; detaillierte bibliografische Daten sind im Internet über <https://dnb.de> abrufbar.

ISBN: 978-3-96729-211-4

Zugl.: Hamburg, Techn. Univ., Diss., 2023

Dissertation, Technische Universität Hamburg

Choriner Str. 85 - 10119 Berlin
verlag@menschundbuch.de – www.menschundbuch.de

Acknowledgement

When I first worked on this topic in my master thesis, I could not anticipate what a wonderful and instructive journey this would be. My deepest gratitude goes to Prof. Dr.-Ing. Arne F. Jacob for giving me this great opportunity to work on this fascinating field of work at his institute. His guidance, encouragement, and support during this whole time is invaluable. Not only did he proofread all of my papers with amazing grasp, he also invested many of his precious evening hours discussing measurement methods with me.

I thank Prof. Dr.-Ing. Christian Damm for his second expertise and Prof. Dr.-Ing. Hoc Khiem Trieu for chairing the examination committee.

I am deeply grateful to Dr.-Ing. Nora Meyne for introducing me to this field of research and guiding me through my master thesis. She encouraged my own ideas and always gave the perfect amount of much needed constructive criticisms. Eventually, these ideas turned into this doctoral thesis.

My thanks also extends to Anja-Maria Doobe-Jöstingmeier, Carmen Hajunga, and Jürgen Winkelmann for supporting me in creating many challenging structures with remarkable precision for my experiments.

Special thanks goes to all of my colleagues and friends at the Institute of High-Frequency Technology for the best working environment. They are able to create joy in every situation and managed to endure my taste in music during our many amazing coffee breaks. I also appreciate our small meetings and the great time we had reading the green physics book in our seminar room.

Last but not least, I would like to thank my family for their unconditional support and encouragement even though they still do not quite understand what I do.

Hamburg, April 2023 ALEKSANDAR SAVIĆ

Abstract

This work aims for single cell characterization. For this purpose, a two-step calibration technique is developed. It is optimized for various technologies like open-ended coaxial lines, printed circuit boards, and microtechnology. In principle, it is based on impedance measurements with a single reference liquid for calibration and is capable of extracting the broadband complex permittivity. The advantages are simplicity, robustness, and scalability down to low volumes. All of the sensors for single cell measurements developed in this work are evaluated with this method. As widely done in literature for cell measurements, microfluidic sensors are developed and produced with microtechnology. Two of the sensors are planar with different geometries. One is a one-port design with three electrodes, the other a two-port version with two electrodes. The electrodes of both versions have a physical contact with one side of the cell. In combination with a mechanical cell trap, they are shown to be capable of measuring single cells and extracting their permittivity. When measuring different cell lines and pathological states, different permittivity footprints are discovered. Measured are living and dead cells from the CHO and human liver cancer HepG2 cell lines. The living HepG2 cells show a lower permittivity compared to the CHO ones, thus making the cells distinguishable. Dead cells of both lines yield quite similar results, but higher values than the living variants. A novel measurement approach at microwave frequencies is introduced. A simple parallel plate capacitor is expanded with the use of a guard electrode to avoid issues with fringing fields. It relies on a design with high impedances and properly positioned electrodes. While this method is known at lower frequencies, its implementation at microwave frequencies is developed here. Its capabilities are demonstrated in simulation and with a printed circuit experiment as a scaled version. With this design, the cell can be illuminated homogeneously through its center. Only this homogeneous field contributes to the extracted permittivity. A microchip version of the sensor is presented, along with the results of CHO cell measurements.

Contents

1 Introduction

In recent years, microwaves have experienced an increased popularity in medical applications [1]–[4]. Their use can be split into two main areas – therapeutic and diagnostic. The former includes, for example, hyperthermia and electroporation, both of which can potentially destroy cancer tissues and cells [5], [6]. The latter includes imaging and dielectric measurements [7]. As shown in [8], the dielectric properties of healthy and cancerous tissues differ. Such differences can be measured with an open ended coaxial line, which is commonly used to determine the dielectric properties of liquids. Similarly, cell lines also differ in terms of permittivity [9]. The cell metabolism is different for healthy and cancerous calls. Their material composition and, in turn the complex permittivity, should change as well. It is to be noted that the measured results in [9] represent mean values of all the present cells in the tissue or cell suspension, respectively. Single cell characterization could lead to more insights.

For some cancer types, e.g. leukemia and ovarian cancer, only single cells are available for analysis. With the aim of finding differences between healthy, dead, and cancerous cells, much effort has been spent in characterizing them at microwaves [10]–[13]. At lower frequencies the capacitance of the cell membrane acts like an open circuit and dominates the measured values. Hence the choice to use microwaves, at which the cell inside has the dominant influence. Additionally, microwaves are non-destructive, at least up to a certain electric field strength, and label-free. Also, the rather involved effect of electrode polarization [14], which occurs in the kHz frequency range and below, is avoided.

Single cell analysis is of interest for two purposes. First, in cancer research there is need for reliable early diagnostic methods. One example is the ovarian cancer with over two hundred thousand new cases each year. Due to its high aggressiveness, the five year death rate is larger than 50 % [15], [16]. In case of breast cancer, mammograms executed yearly for 10 consecutive years yield a false positive rate is about 95 % [17]. The second purpose is the quality control in the production of biopharmaceuticals with cell lines like Chinese Hamster Ovary (CHO) [18], [19]. In the long production cycles, in the order of several months, extensive quality control is necessary to ensure that the conditions are not altered.

In order to pursue to above mentioned goals, new calibration methods, sensors, and broadband permittivity results at microwave frequencies are required. This work provides all that for measurements on single biological cells.

Structure of this work

Chapter 2 introduces some theoretical basics regarding permittivity, dispersion and its modeling, and state-of-the-art sensors. Different permittivity measurement methods are elaborated and narrowed down for the purpose of single cell characterization.

Chapter 3 focuses on the calibration method for accurate permittivity extraction of small samples such as single biological cells. As often done in literature, the calibration is first developed, optimized, and verified on liquids with scaled sensors. Subsequently experiments with a microchip sensor are carried out with Cole-Cole fitting and an uncertainty analysis for the developed procedure.

The developed planar microchip sensor designs are presented in Chapter 4, alongside with single cell broadband permittivity measurement results, extracted Cole-Cole parameters, and comparison to literature.

Planar sensors measure the cells from one side. In order to measure the middle part of the cell there is no way around two opposite metal plates, i.e. a parallel plate capacitor. Due to the irregular cell geometry, the contribution of stray fields introduces further calibration issues which need to be handled properly. Chapter 5 addresses this issues and suggests a novel electrode configuration and equivalent circuit evaluation. Proof-of-principle measurements using a scaled sensor are made. Measurements of single cells with microchip sensors are also provided.

Theoretical Basics 2

The characterization of cells in this work is done in terms of complex relative permittivity, a material property describing the change of the electric field relative to vacuum. Permittivity, in general, is of dispersive nature. Frequency ranges in which the permittivity significantly changes due to a certain effect, i.e., a dispersion mechanism, are usually of high interest. These so called relaxation processes can be modeled with equations, e.g., Debye or Cole-Cole model [20], which makes the differentiation between materials straightforward.

For biological cells, the term effective relative permittivity is used in this work. In literature it is sometimes noted as $\underline{\epsilon}_{\mathrm{eff,r}}$, which results from $\underline{\epsilon}_{\mathrm{eff}} = \underline{\epsilon}_{\mathrm{eff,r}} \cdot \epsilon_0$. It describes the cell as a whole structure rather than single materials within the cell. For this reason, biological cells have several relaxation processes as will be described in the following section. With the effective permittivity, an equivalent circuit representation with lumped elements can be derived and vice versa. For the course of this work, the complex values will be written with an underscore, e.g., $\underline{\epsilon}_{\mathrm{r}} = \epsilon_{\mathrm{r}}' - \mathrm{j}\epsilon_{\mathrm{r}}''$.

2.1 Dispersion Mechanisms in Biological Cells

Biological cells experience several relaxation processes [21], [22]. The first one, the α-dispersion, is due to the counterion effects and occurs below 10 kHz. Up to roughly 100 MHz the β-dispersion due to Maxwell-Wagner effects occurs [21]. Because of their high water content, it is to be expected that biological cells experience the γ-dispersion in the GHz range, similarly to water. This dispersion is the focus of ongoing research and of this work. Some authors [23] mention a δ-dispersion in the range between 0.1 GHz and 3 GHz. Its interpretation is somewhat difficult due to the absence of a (single) dominant mechanism. The four mentioned dispersion cases are exemplarily plotted in Figure 2.1. Further relaxation processes exist, e.g., atomic ones at THz frequencies, which are not of interest for the current work.

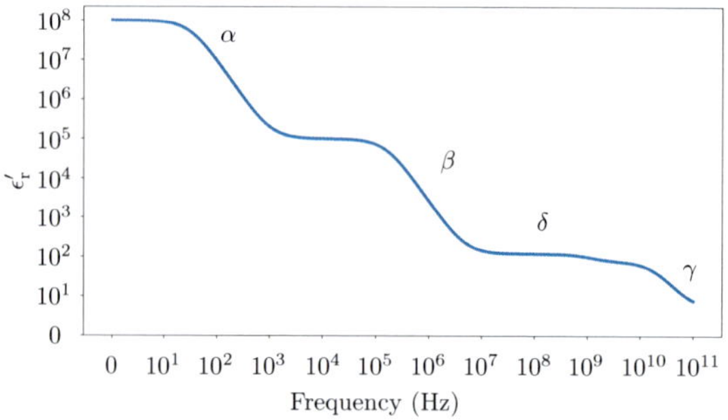

Figure 2.1: Exemplary real part of the permittivity spectrum of a biological cell versus frequency in a logarithmic scale.

α-Dispersion

Applying an outer electric field influences the ion clouds on both the inner and the outer side of the membrane. An alternating electric field moves the ions, which are bound to the membrane surface. With increasing frequency, the ions cannot follow the applied field anymore. Thus, the opposing field, and with it the permittivity, are reduced. This effect is called the α-dispersion and occurs below 10 kHz [24].

The intensity of the α-dispersion is not only dependent on the cell itself, but also on the external medium. A culture media containing biological cells is usually an ionic solution. The measurements of α-dispersion are somewhat difficult because of the interference with an another effect, namely the electrode polarization. At low frequencies, this effect introduces the so called double layers in the electrodes vicinity, which, in turn, influence the measured impedance. An inclusion of the double layer capacitance into the evaluation is necessary, for example by using the Helmholtz or the Gouy-Chapman model. Additionally, sorption and faradaic currents would need to be considered [21].

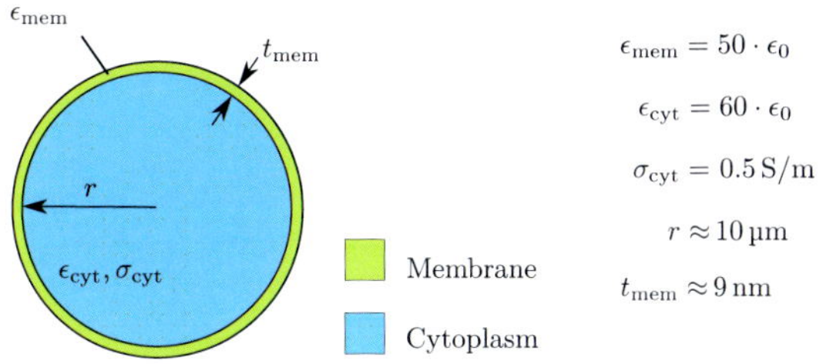

Figure 2.2: A possible low frequency cell representation in simulation.

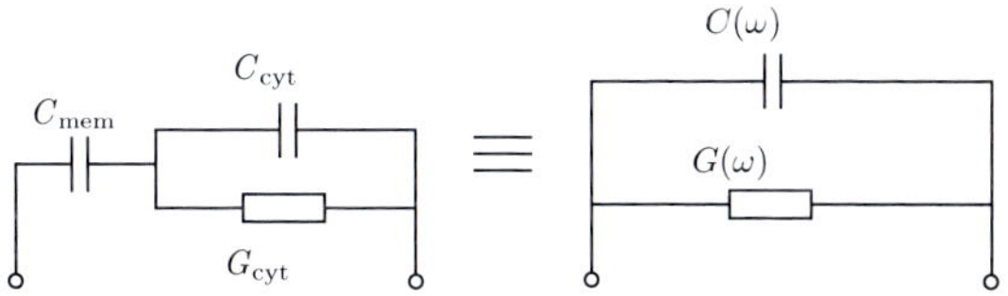

Figure 2.3: A possible equivalent circuit representation of a single cell between two opposing electrodes. Alternatively, a frequency dependent total capacitance $C(\omega)$ and conductance $G(\omega)$ may be used.

β-Dispersion

Between roughly 10 kHz and 100 MHz a biological cell can be represented as a sphere with a thin membrane as depicted in Figure 2.2 [25]. The cytoplasm is represented as a single material here, but is non-homogeneous in reality. A possible equivalent circuit representation in case of two opposing electrodes is depicted in Figure 2.3. The three lumped elements, namely the membrane capacitance C_{mem}, the cytoplasm capacitance C_{cyt}, and conductance G_{cyt}, can be summarized in a single RC-circuit. Optionally, the membrane conductance G_{mem} can be included, but is often omitted because it is negligible compared to C_{mem}. The resulting capacitance $C(\omega)$ and conductance $G(\omega)$ are frequency dependent and directly proportional to the real and imaginary parts of the resulting effective permittivity $\underline{\epsilon}_{eff}$, respectively. With

$$\underline{\epsilon}_{eff}(\omega) = C_m r \left[\frac{j\omega\tau_c + 1}{j\omega(\tau_m + \tau_c) + 1} \right], \tag{2.1}$$

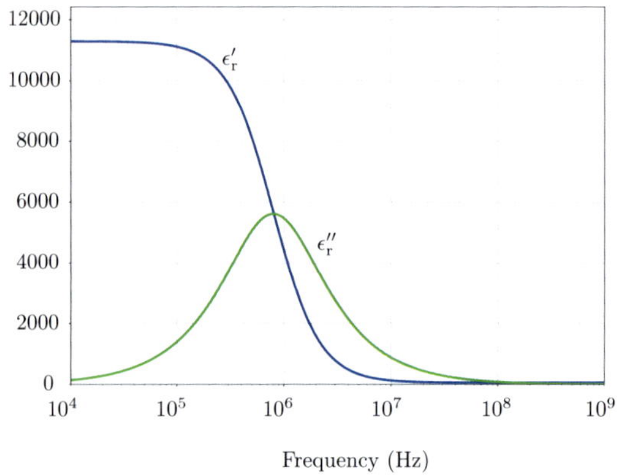

Figure 2.4: Exemplary effective complex permittivity of a biological cell for the frequency range in which the β-dispersion occurs.

the effective permittivity $\underline{\epsilon}_{\text{eff}}$ can be calculated [26]. It depends on the membrane capacitance per unit area, which is $C_{\text{m}} = \epsilon_{\text{mem}}/t_{\text{mem}}$, and on the time constants, which are $\tau_{\text{c}} = \epsilon_{\text{cyt}}/\sigma_{\text{cyt}}$ and $\tau_{\text{m}} = C_{\text{m}} r/\sigma_{\text{cyt}}$. Typical values are $C_{\text{m}} = 1\,\mu\text{F/cm}^2$, $\sigma_{\text{cyt}} = 0.5\,\text{S/m}$, $\epsilon_{\text{mem}} = 10\,\epsilon_0$, and $\epsilon_{\text{cyt}} = 60\,\epsilon_0$.

In the mentioned frequency range, the relative $\underline{\epsilon}_{\text{eff,r}}$ can be in the order of several thousands due to the dominance of the membrane capacitance. The information about the cell cytoplasm is limited. An exemplary plot of $\underline{\epsilon}_{\text{eff,r}}$ for a biological cell is shown in Figure 2.4.

γ-Dispersion

The γ-dispersion, on which this work focuses mostly, occurs in the GHz range. It is caused by the dipolar mechanism in polar media such as water [20]. Up to several GHz the orientation of dipoles in water can follow the applied alternating electric field. Eventually this ceases to be the case with increasing frequency causing friction loss, which is usually called dielectric loss. This results in a continuous change of the real and imaginary parts of the permittivity up to about 100 GHz.

Figure 2.5 of the permittivity spectrum versus frequency of a 0.9% NaCl solution displays the γ-dispersion [27]. Due to the existing DC-conductivity in NaCl solutions and the relation $\epsilon_r'' \propto \frac{\sigma_{DC}}{\omega}$, the ϵ_r'' curve has a pole at DC. The dielectric loss of the γ-dispersion comes on top of the ohmic loss. With the ϵ_r'' representation, it can be seen that around several GHz the dominating loss mechanism changes from ohmic to dielectric. To calculate the total conductivity an integration over ω is necessary. If desired, the measured impedance of materials with γ-dispersion can be represented by an equivalent circuit as sketched in Figure 2.6, where G_{DC} stands for the DC-conductivity, C_∞ is the capacitance at frequencies beyond the dispersion region, and C_D and G_D stand for the dispersion mechanism [25], [28].

Biological cells consist up to about 70-90% of water, which is why this dispersion is of high interest. Also, the culture media for commercially available CHO cells are similar to a 0.5% NaCl solution. Optionally, the culture media can be exchanged with a different fluid, e.g., phosphate buffered saline (PBS), which is more like a 0.9% NaCl solution.

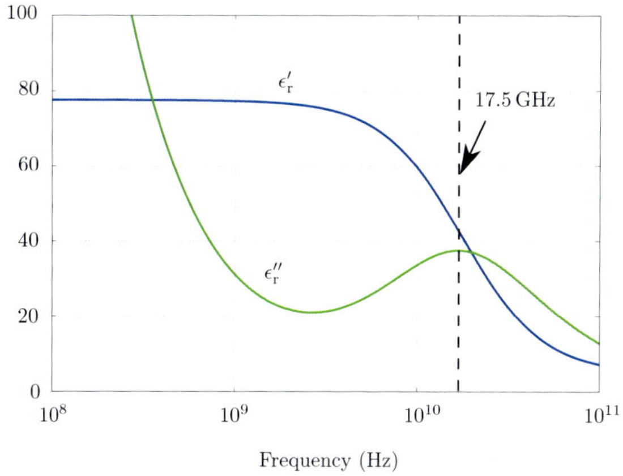

Figure 2.5: Permittivity spectrum of a 0.9% NaCl solution.

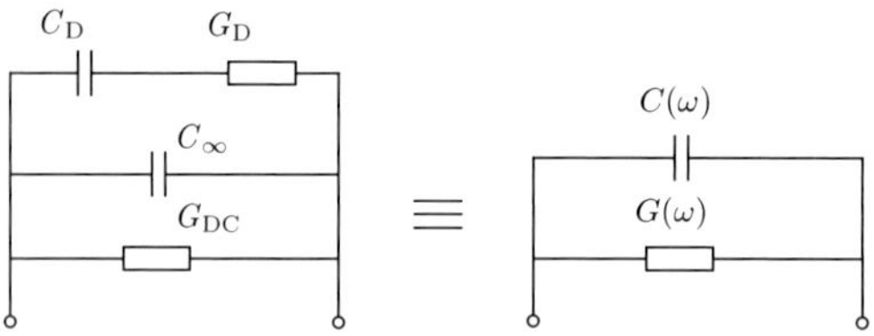

Figure 2.6: A possible equivalent circuit representation of the γ-dispersion.

2.2 Mathematical Formulations of Dispersion Processes

In permittivity studies, authors commonly provide model parameters of the measured fluids after some fitting is applied, which is often root mean square. This allows easy reconstruction of the permittivity curves and comparison of different works.

Models for a Single Relaxation Process

The most famous and most used models for single relaxation processes are the classical Debye model and the Cole-Cole model, which is basically a modified version of the former one to account for different spectral distributions [20]. Depending on α, formulation

$$\underline{\epsilon}_{\mathrm{r}}(\omega) = \epsilon_\infty + \frac{\epsilon_{\mathrm{static}} - \epsilon_\infty}{1 + (\mathrm{j}\omega\tau)^{1-\alpha}} + \frac{\sigma_{\mathrm{DC}}}{\mathrm{j}\omega\epsilon_0} \tag{2.2}$$

either represents a Debye model ($\alpha = 0$) or a Cole-Cole model ($\alpha \neq 0$). The permittivity limits are given by $\epsilon_{\mathrm{static}}$ and ϵ_∞, the DC-conductivity by σ_{DC}, and the relaxation time constant by τ. The factor α models the previously mentioned spectral distribution.

Saline solutions are easy to produce and are well studied in literature. The variations after repeated measurements are low, in the order of $\Delta\epsilon_{\mathrm{r}} = \pm 0.2$, especially at low concentrations. Thus, they are of high interest to be used as calibrations standards and test fluids. Table 2.1 lists the Cole-Cole model parameters at room temperature for the most used saline solutions in this work. Higher concentrations lead to more loss and a larger $\tan\delta = \frac{\epsilon_{\mathrm{r}}''}{\epsilon_{\mathrm{r}}'}$, which results in more variations after repeated measurements. The reason for this effect is not perfectly clear but possibly depends on the measurement method and the influence of increasing conductivity on the applied models and calibration procedures.

Table 2.1: Cole-Cole model permittivity parameters for various saline solutions measured at 21°C [27].

Fluid	ϵ_{static}	ϵ_∞	τ	α	σ_{DC}
NaCl 0.5%	78.66	5	9.15 ps	0.005	0.87 S/m
NaCl 0.9%	77.64	5	9.10 ps	0.00695	1.45 S/m
NaCl 1.0%	77.43	5	9.084 ps	0.0074	1.59 S/m
NaCl 1.5%	76.52	5	9.04 ps	0.01	2.24 S/m
NaCl 2.0%	75.71	5	8.99 ps	0.0134	2.89 S/m
NaCl 2.5%	74.88	5	8.93 ps	0.0172	3.55 S/m
NaCl 3.0%	74.02	5	8.87 ps	0.02	4.23 S/m

The permittivity is, of course, temperature dependent. An increase in temperature results in an increase in the DC-conductivity, decrease in the static permittivity, and a slight change of the relaxation time constant τ as shown in Table 2.2 for the temperature of 36°C.

These literature results are to be handled with care as they are not necessarily extracted from broadband measurements over the whole frequency range of the γ-dispersion. In [27], open-ended coaxial line (OEC) measurements up to 20 GHz are used, while the values in [29] are extracted by using several single frequencies and temperatures. Often, some interpolation is involved, which is why, for example, the presented ϵ_∞ might be inaccurate.

Table 2.2: Debye model permittivity parameters for various saline solutions at 36°C [29].

Fluid	ϵ_{static}	ϵ_{∞}	τ	σ_{DC}
NaCl 0.5%	73.02	4.9	9.98 ps	1.07 S/m
NaCl 0.9%	71.78	4.9	9.98 ps	1.89 S/m
NaCl 1.0%	71.48	4.9	9.97 ps	2.07 S/m
NaCl 1.5%	69.97	4.9	9.97 ps	3.09 S/m

Debye Model for Multiple Relaxations

For a mixture of, e.g., two different fluids, both of their respective dispersion processes influence the effective permittivity. This effect can be described by

$$\underline{\epsilon}_{\mathrm{r}}(\omega) = \epsilon_{\infty} + \frac{\Delta\epsilon_1}{1 + \mathrm{j}\omega\tau_1} + \frac{\Delta\epsilon_2}{1 + \mathrm{j}\omega\tau_2} \tag{2.3}$$

which contains several relaxation time constants and $\Delta\epsilon_{\mathrm{r}}$-terms. The latter represent the permittivity differences between the steps of the relaxation processes. An additional DC-conductivity and the factor α could be added to the equation.

Table 2.3 contains the parameters for two water–ethanol mixtures with different ethanol mole fractions.

Table 2.3: Debye model permittivity parameters for water-ethanol mixtures at 25°C [30].

Ethanol mole fraction	$\Delta\epsilon_1$	$\Delta\epsilon_2$	ϵ_{∞}	τ_1	τ_2
0.22	44	6.1	4.3	33 ps	8 ps
0.36	36.1	3.8	4.5	36.1 ps	9 ps

2.3 Permittivity Measurement Techniques

This section briefly introduces the most used measurement techniques for microwave permittivity [25], [31]–[33]. All of them offer the possibility to extract the real and imaginary parts of the permittivity from the scattering parameters. The imaginary part is extracted either directly or via the loss tangent $\tan(\delta)$, which is basically a ratio of the real and imaginary parts. Numerous variations of the techniques themselves or their calibration procedures exist. A brief overview of the different methods is given in Table 2.4.

Table 2.4: Overview of permittivity measurement techniques.

Method	Advantages	Disadvantages
Free space	- high frequencies - non-destructive	- flat materials needed - multiple reflections limit accuracy
Reflection	- broadband - non-destructive	- electrically large sample size - flat surface needed in case of solids
Transmission line	- simple - broadband	- limited geometry - not suitable for single cells
Resonance	- sensitivity - high accuracy	- only single/few frequencies - possibly complex analysis
Impedance	- broadband - small samples	- specially designed electrodes required - limited maximum frequency

Free space methods are intended for electrically large, flat objects where direct contact of a sensor and object might be unpractical. Usually, two antennas around the sample are required to extract the permittivity from the measured reflected and transmitted waves. This method can be employed in hostile environments and allows for non-destructive measurements. The accuracy of the results, however, is limited due to effects such as diffraction, multiple reflections, and surface roughness. Recent advances can be found in [34], [35].

Reflection based permittivity measurements can be carried out with an OEC [36]–[39]. This is often done to characterize fluids as the method is fairly accurate, broadband, and simple. The OEC is inserted into a sufficiently large container to avoid multiple reflections. Ideally, there are no gaps or air bubbles between the OEC and the sample. At least three known standards are necessary for calibration. The permittivity information is extracted from the reflection coefficient at the interface between the OEC and the sample. An equivalent circuit analysis of the OEC with the calibration technique developed in this work is also possible, as will be demonstrated in Section 3.3.

Transmission line [40]–[43] measurements for permittivity extraction are popular for electrically large samples which can in some way be integrated into a transmission line. Ideally, this is done at the locations of high field intensity. A solid sample could be inserted into a rectangular waveguide or a coaxial line. A coplanar waveguide (CPW) in combination with microfluidics allows for fluid measurements as well [44], [45]. The core of this method is the determination of the propagation constant γ, for which different procedures exist. For electrically small samples the determination of γ and, in turn, of the permittivity becomes less accurate. Additionally, air gaps and unintended discontinuities can negatively influence the results.

Resonance methods [46]–[49] are highly sensitive and often employed in the permittivity and loss determination of low loss materials for printed circuit board (PCB) technology. The material to be characterized needs to be placed near a field maximum of the resonator. This influences the resonant frequency and the quality factor as measured by the scattering parameters. A drawback is the limited number of frequency points at which the measurements occur. This can only be overcome by employing multiple different resonators which is more involved than the previously mentioned methods. Additionally, the network analyzer needs to have a high frequency resolution to achieve the high accuracy. As an advantage, the sample size is less of an issue as long as the field is highly concentrated by means, e.g., of small metallic tips.

Impedance measurements [50], [51] involve electrically small samples which are to be represented by lumped elements in an equivalent circuit. This is fairly trivial in the low frequency range where the complex permittivity can be extracted with a simple parallel plate capacitor designed with minimal fringing field influence. In the microwave range this can be done by setting the reference plane close to the sample or by subsequent de-embedding procedures. Some parasitic effects might occur and need to be dealt with accordingly, e.g., by specially designing the electrodes for the sample.

2.4 State-of-the-art Cell Measurements

Transmission line based sensors are suitable for measurements of cell suspensions. In [52] human umbilical vein endothelial cells (HUVEC) are measured in a suspension and a permittivity contrast compared to a medium without cells is demonstrated. DOHH-2 cell suspensions measured in [53] show different permittivity results with and without treatment with saponin. Saponin leads to holes in the cell membrane and leads to death of cells. Differences in terms of permittivity for different B lymphoma cells are discovered in [1]. Different pathological states of CHO cells in suspensions also lead to a permittivity contrast as measured in [54] and [45].

The number of cells exposed to the electric field, as well as the volume portion compared to the medium is hard to estimate. Extracted permittivity represents a mean value of all of the measured cells and the surrounding medium. Recent advances of this method with microchip technology can be found in [55], extending the maximum frequency to 110 GHz.

Resonance based sensors are suitable for measurements of cell suspensions as well as single biological cells. Different cell lines in suspension have been distinguished based on their permittivity with this method in [10] and [9]. In [56]–[58], CHO cells in suspensions with different pathological states have been differentiated. A contrast between single CHO cells and the surrounding medium is measured in [2]. A difficulty in this approach is concentrating the electric filed so that the electric does not penetrate into the surrounding fluid. Due to too wide electrodes in [2], the extracted permittivity is a mixture of the cell's and the medium's properties.

Impedance based sensors are appropriate for broadband measurements of both cell suspension and single cells. The former is measured in [59], with results varying with cell concentration. Measurement results of single B lymphoma cells [60], and single THP1 cells [12], [61], [62] are available in literature for frequencies up to 40 GHz. The result representation is in form of capacitive difference ΔC and conductance difference ΔG. Partially, the material property, i.e. the permittivity, is presented at certain frequency points [61] after calibration by simulation results. A different approach is presented in [3], [13], [63], [64] where Jurkat cells are measured. The frequency ranges from kHz up to 20 GHz. The results are presented in form of the difference in the scattering parameters and equivalent circuit representations of cells. This method is often employed at frequencies in the kHz and MHz range, e.g., in [65]–[67].

The presented single cell results are often dependent on cell size and sensor dimensions. This makes a comparison of the sensors and cells rather difficult. Results regarding the material property of single cells, i.e. the permittivity, are scarce and only available for single frequency points. The goal of this work is to present a simple calibration method for obtaining broadband permittivity results of single cells, alongside with measurements and comparisons between different sensors and cell lines.

3 Calibration Procedure Development

Near-field dielectroscopy is suitable for single cell characterization due to their small size in the order of 10 µm. The main aim is to develop a procedure which is simple, robust, easily applicable, requires few reference materials, and allows rapid re-calibration. The latter is advantageous because cells have a limited life span after they are prepared for measurements.

In this work, a broadband method based on impedance measurements, is chosen to obtain as many data as possible. The aim is to provide enough data for a statistical analysis and parameter fitting into a Cole-Cole model.

In this chapter, a two-step calibration procedure based on equivalent circuits is developed. Here, only one reference fluid is required, in addition to the standard transmission-reflection-line (TRL) calibration. The applicability of the calibration procedure on the classical OEC measurement solution is demonstrated. The procedure is validated in PCB technology with different fluids in a one-port scenario as a cost effective alternative to microchips. Optimization steps in regard to the accuracy and applicable frequency range are carried out. An exemplary microchip is used to validate the calibration procedure in micro-technology and to extend it for a two-port measuring scenario. This is essential as sensors for single cells are usually made with micro-technology. A Cole-Cole parameter fitting procedure is tested and the measurement results are compared to typical literature values.

This chapter is based on publications [25], [68]–[70]. In [25], the basic calibration concept is provided along with a practical example in PCB technology. There, simulation results of a theoretical microchip sensor are also presented. Here, the procedure is optimized to reach higher frequencies with minimal effort. To this end, the PCB sensor from [25] is optimized with two additional versions and the theoretical considerations regarding the microchip implementation are experimentally verified. For the sake of completeness, the basic calibration concept and the first PCB sensor from [25] are briefly summarized here.

Simulation is executed with Ansys HFSS and all measurements are performed at room temperature of about 21 °C.

3.1 Equivalent Circuit Model

Impedance sensors for lossless or low-loss materials are essentially capacitive and can be modeled by a single capacitor at low frequencies. A single lumped element is usually not sufficient at microwave frequencies. Thus, additional elements are needed.

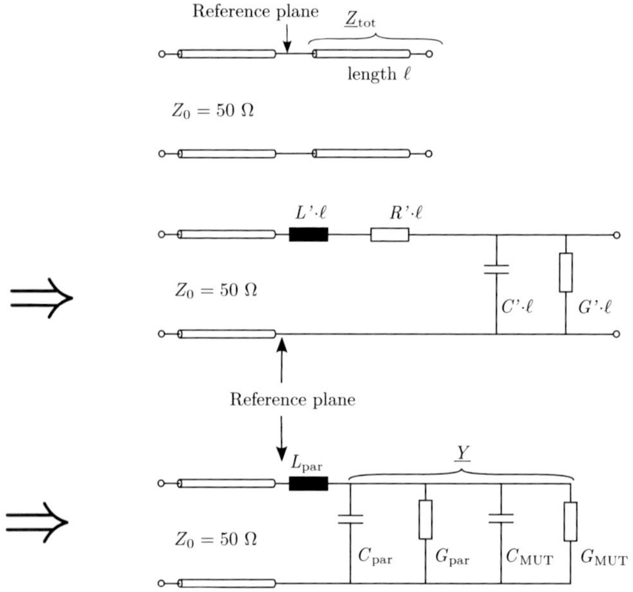

Figure 3.1: Equivalent circuit of a short transmission line, modified to represent a one-port impedance sensor.

A simple, electrically short transmission line has several possible lumped element representations. One of these is depicted in Figure 3.1 and is valid for lengths of approximately $\ell < \frac{\lambda}{10}$, with λ being the wavelength. Beyond this limit, several such circuits need to be cascaded. The $R'L'G'C'$ transmission line parameters stand for the ohmic loss, self inductance, dielectric loss, and the capacitance per unit length [41]. The characteristic wave impedance of the transmission line is denoted with Z_0 and is typically 50 Ω. It can also be a complex value, e.g., when the thickness of a transmission line is in the order of the skin depth. In any case, an accurate Z_0 is required to extract the equivalent circuit behind the reference plane.

Since this representation basically includes all the important effects between two metal structures, it is modified for the purpose of calibrating impedance sensors. The total impedance $\underline{Z}_{\text{tot}}$ is easily extracted from the simulated or measured reflection coefficient $\underline{r}$ according to

$$\underline{Z}_{\text{tot}} = \frac{Z_0 \cdot (1 + \underline{r})}{1 - \underline{r}}. \tag{3.1}$$

Figure 3.1 also shows the modified equivalent circuit. The ohmic loss is neglected as it is usually small compared to the other elements. The capacitance and the conductance are each split into two parts representing parasitic elements (labeled 'par') and the material under test (MUT). The capacitance is basically a product of the real part of the permittivity and a geometry dependent factor in meters. Similarly, the conductance depends on the conductivity, which is $\sigma = \omega\epsilon_0\epsilon_{\text{r}}''$, and also a geometrical factor. This yields the equations

$$C_{\text{MUT}} = \epsilon_0 \cdot \epsilon_{\text{r,MUT}}' \cdot \left(\frac{A}{l}\right)_{\text{MUT}} \tag{3.2}$$

and

$$G_{\text{MUT}} = \omega\epsilon_0 \cdot \epsilon_{\text{r,MUT}}'' \cdot \left(\frac{A}{l}\right)_{\text{MUT}}, \tag{3.3}$$

where $\left(\frac{A}{l}\right)_{\text{MUT}}$ represents the geometry dependent factor and depends on the electric field distribution. Its nomenclature is chosen on the basis of the equations for a parallel plate capacitor. There, due to the simple field distribution, the surface area A and the length l can simply be determined from the geometry of the parallel plate arrangement. For two arbitrary metal structures, the field distribution can be complex. Therefore, simulation or experimental data is required to determine this factor.

Here, the admittance is represented without the parasitic inductance according to

$$\underline{Y} = G + j\omega C = \frac{1}{\underline{Z}_{\text{tot}} - j\omega L_{\text{par}}} \tag{3.4}$$

with

$$C = C_{\text{par}} + C_{\text{MUT}} = C_{\text{par}} + \epsilon_0 \cdot \epsilon'_{\text{r,MUT}} \cdot \left(\frac{A}{l}\right)_{\text{MUT}} \tag{3.5}$$

and

$$G = G_{\text{par}} + G_{\text{MUT}} = G_{\text{par}} + \omega\epsilon_0 \cdot \epsilon''_{\text{r,MUT}} \cdot \left(\frac{A}{l}\right)_{\text{MUT}}. \tag{3.6}$$

Both C and G are related to the same electric field distribution. Therefore, the factor $\left(\frac{A}{l}\right)_{\text{MUT}}$ is the same for both. Depending on the frequency, the value G_{par} might be negligible just like the ohmic losses. Assuming that the parasitic values do not change when exchanging the MUT, the geometric factor is given by

$$\left(\frac{A}{l}\right)_{\text{MUT}} = \frac{1}{\epsilon_0} \cdot \frac{dC}{d\epsilon'_{\text{r,MUT}}} \tag{3.7}$$

and

$$\left(\frac{A}{l}\right)_{\text{MUT}} = \frac{1}{\omega\epsilon_0} \cdot \frac{dG}{d\epsilon''_{\text{r,MUT}}}. \tag{3.8}$$

The value $\left(\frac{A}{l}\right)_{\text{MUT}}$ is a slope for the upcoming calibration curves and basically represents a measure for the sensitivity of the setup.

3.2 Calibration Concept

3.2.1 Basic Principle

For classical reflection based permittivity measurements, e.g., OEC, three known standards are necessary, usually a short, an open and a reference material [36], [71]. A short can be realized with mercury [37]. These three standards are not practical in the present context due to the small size of the MUT, i.e., biological cells, and of the sensors involved.

Recently, methods with two [72], three [73], and five [74] reference liquids have been developed. These methods are rather involved in terms of time required to calibrate and re-calibrate sensors. Biological cells in culture have a limited measurement window before they die and change properties. In [75], [76] a linear relationship is found between the real and imaginary parts of the permittivity and the capacitance and conductance, respectively. However, this is demonstrated only in simulation up to about 1 GHz. As will be shown later, the proposed calibration procedure achieves higher frequencies in the order of 20-30 GHz both in simulation and measurements.

The developed two-step calibration procedure consists of a standard TRL calibration to set the reference plane and two reference measurements. According to Equations 3.4, 3.5, and 3.6, the unknown calibration parameters are L_{par}, C_{par}, G_{par}, and $\left(\frac{A}{l}\right)_{\mathrm{MUT}}$. Four degrees of freedom are necessary to determine these factors. To this end, the values C and G are measured, while the MUT properties $\epsilon'_{\mathrm{r,MUT}}$ and $\epsilon''_{\mathrm{r,MUT}}$ are varied. Only two complex measurements of known materials are necessary to calculate the required parameters. An obvious choice for one calibration material is air. Because saline solution exhibits a significantly different permittivity, it is suitable as a second calibration material. With these choices, the procedure is a single reference liquid calibration. Whereas the Equations 3.5 and 3.6 are sufficient to extract C_{par}, G_{par}, and $\left(\frac{A}{l}\right)_{\mathrm{MUT}}$, the extraction of L_{par} is not as straightforward. This is elaborated in the following section.

3.2.2 Parasitic Inductance

As demonstrated in [68], an inaccurate inductance value limits the upper operating frequency of the sensor. Thus, L_{par} needs to be known as accurately as possible.

Its extraction directly with the presented equations has shown to be involved and inconsistent. Additionally, a direct measurement of this inductance is obstructed by the experimental inaccuracies and the dispersion of the materials. The latter results in frequency dependent lumped elements in the equivalent circuit representation. Another possibility is to use simulation results but this might lead to inaccuracies due to production tolerances.

Here, an alternative approach is presented to improve the sensor performance. The slope $\left(\frac{A}{l}\right)_{\mathrm{MUT}}$ is slightly frequency dependent because the equivalent circuit is an approximation and Z_0 is frequency dependent as well. Its value $\left(\frac{A}{l}\right)_{\mathrm{MUT}}$ is equal for C

and G, as demonstrated later in Section 3.4.2, and the condition for a single frequency point yields

$$\frac{1}{\epsilon_0} \cdot \frac{dC}{d\epsilon'_{\mathrm{r,MUT}}} = \frac{1}{\omega\epsilon_0} \cdot \frac{dG}{d\epsilon''_{\mathrm{r,MUT}}}. \tag{3.9}$$

Extracting Y with an inaccurate inductance inevitably leads to different slopes for C and G over the considered frequency range. To optimize the sensor performance the condition

$$\min \sum_f \left| \left(\frac{A}{l}\right)_{\text{from C}} - \left(\frac{A}{l}\right)_{\text{from G}} \right|^2 \tag{3.10}$$

is imposed [70]. The upper frequency limit of the sensor could be determined by means of simulation. This approach to extract L_{par} is demonstrated with the PCB sensor in Section 3.4.3. Alternatively, the inductance could be estimated with the inductance per unit length L' of the transmission line. This would be quicker but less accurate than the proposed solution.

3.2.3 Complex Characteristic Impedance

Another matter to consider is the case of a complex characteristic impedance, which usually occurs in microchips with thin metal layers. A possible method is presented in [77], [78], for which a modified version is outlined here and demonstrated in Section 3.5.

The typical metalization thickness in microchips is several hundreds nanometer to about a micrometer. This results in a significant imaginary part of $\underline{Z}_0$, especially in the lower GHz range, which influences the impedance extraction of the sensor. An inaccurate $\arg(\underline{Z}_0)$ introduces cross dependencies between C and G, which are highly undesired. In [77] a method of determining $\underline{Z}_0$ with the complex propagation constant $\underline{\gamma}$ is proposed. Here, the same method is applied to approximate $\arg(\underline{Z}_0)$. The magnitude of $\underline{Z}_0$ is a scaling factor and hence not necessary for the calibration procedure. Using the $R'L'G'C'$ values per unit length, the relation between $\underline{Z}_0$ and $\underline{\gamma}$ is

$$\frac{\underline{\gamma}}{\underline{Z}_0} = G' + \mathrm{j}\omega C'. \tag{3.11}$$

For low loss substrate materials, as used in this work, typical values per unit length for a CPW in microtechnology are in the order of $\omega C' = 850\,\mathrm{mS/m}$ and $G' = 5\,\mathrm{mS/m}$. This G' is negligible [70] and the angle of $\underline{Z}_0$ can be approximated as

$$\arg(\underline{Z}_0) = \arg\left(\frac{\underline{\gamma}}{G' + \mathrm{j}\omega C'}\right) \approx \arg(\underline{\gamma}) - 90°. \tag{3.12}$$

The typical error due to this approximation is about 1°. An elementary S-parameter measurement of the line section after TRL calibration is adequate to extract $\arg(\underline{\gamma})$ because $\underline{S}_{11} = 0$ and $\underline{S}_{12} = e^{\underline{\gamma} l}$. The length l of the section does not have to be known exactly as $\arg(\underline{\gamma})$=$\arg(\underline{\gamma}$l), which simplifies the evaluation. Thus, the $\arg(\underline{Z}_0)$, if required, can easily be determined experimentally.

3.3 Open-Ended Coaxial Line (OEC) Sensor

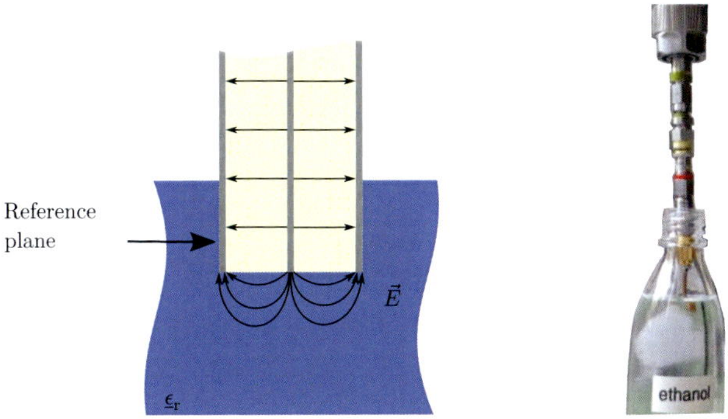

Figure 3.2: An OEC sensor dipped into a liquid.

OEC permittivity measurements have been around for decades [79]. They are intended for liquid volumes down to mililiters. Usually, three or more reference materials are needed for calibration – air, at least one reference liquid, and a short circuit realized by mercury. The advantage of such calibration is that all of the standards are applied to the same structure, hence the reference plane is placed at the intersection of the MUT and the sensor. The disadvantages here are the necessity of several reference liquids and an involved re-calibration process.

Here, the previously developed two-step calibration procedure is applied on a classical OEC permittivity sensor as presented in Figure 3.2. An SMA connector is used, which limits the upper frequency limit to 18 GHz. The employed coaxial line has an outer diameter of 2.3 mm. Ideally, the reference plane is placed at the MUT-sensor intersection, which is a simple task in simulation. In practice, however, due to production and measurement tolerances, this cannot be realized exactly. These tolerances are taken into account by the second calibration step and the previously mentioned parasitic elements C_{par} and G_{par}.

Calibration Curves

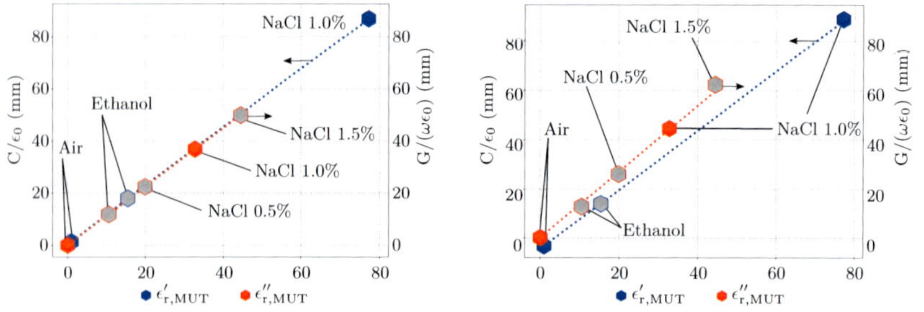

(a) From simulation results. (b) From measurement results.

Figure 3.3: Normalized 1 GHz results for C and G of the OEC sensor for different liquids. Capacitance values of 0.5 %, 1.0 %, and 1.5 % NaCl solutions are similar and thus not all are shown.

The simulated calibration curves of the OEC sensor at 1 GHz are plotted in Figure 3.3(a). Because the reference plane is perfectly set, there are no parasitics, i.e., $C_{\mathrm{par}}=0$ and $G_{\mathrm{par}}=0$. Graphically this means that the curves overlap and, if they are extended, pass through the origin of the coordinate system. The slopes are parallel as expected by the model, their value being $\left(\frac{A}{l}\right)_{\mathrm{MUT}}=1.12\,\mathrm{mm}$.

The measured calibration scales are plotted in Figure 3.3(b). They are approximately parallel but do not overlap as in simulation. The value C_{par} is not zero, indicating that the reference plane is not perfectly set as intended. The slope here is $\left(\frac{A}{l}\right)_{\mathrm{MUT}}=1.2\,\mathrm{mm}$.

Permittivity Results

The simulation permittivity results are presented in Figure 3.4. The dotted lines, i.e., the input data, represent the materials definitions in the simulation, as taken from literature. The values extracted with the S-parameters are in good agreement with the input data. The errors at 10 GHz are below 1% for the most part.

Despite the practical issue of an imperfect location of the reference plane, the measurement results in Figure 3.5 are in good agreement with literature values. This shows that the proposed calibration procedure is capable of extracting the complex permittivity. Higher frequencies could be reached by using a coaxial line with a smaller diameter and appropriate connectors.

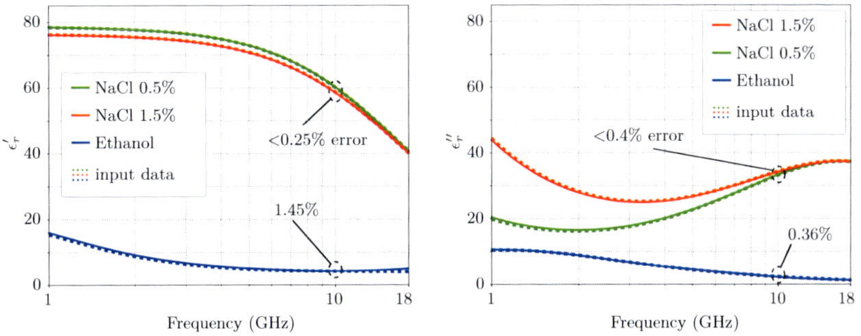

(a) Real part of the permittivity. (b) Imaginary part of the permittivity.

Figure 3.4: Permittivity values extracted from simulation results versus frequency for the OEC sensor.

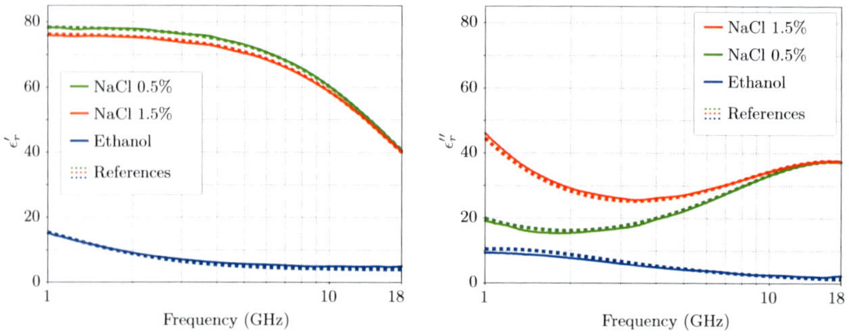

(a) Real part of the permittivity.

(b) Imaginary part of the permittivity.

Figure 3.5: Permittivity values extracted from measurements versus frequency for the OEC sensor.

3.4 Printed Circuit Board (PCB) Sensors

To validate the previously outlined calibration procedure, an inexpensive PCB technology is used, here. With a manufacturing resolution of about 100 µm, measurements of microliter volumes are possible. Three sensors are developed successively and the upper operating frequency limit is improved with each iteration. This limit is defined as the frequency above which the deviations of the measured material properties from the literature values are $\Delta\epsilon_r = 1$.

All sensors are calibrated with the same materials, namely a 1% NaCl solution with permittivity values from [27] and air. The former is prepared with an accuracy of about 1 mg for 50 ml. This preparation accuracy results in permittivity deviations in the order of $\Delta\epsilon_r' = 0.005$ and $\Delta\epsilon_r'' = 0.05$, which is less than the measurement errors reported in the literature.

For ethanol and methanol the reference values are taken from [30], [80] which, notably, can experience large deviations after repeated measurements in the order of $\Delta\epsilon_r = 1$ compared to about $\Delta\epsilon_r = 0.1$ for saline solutions. This is an additional reason to use saline solutions as calibration standards. The measurement equipment includes a ZVA50 network analyzer from Rohde&Schwarz in combination with either standard SMA connectors or ACP40 measurement probes. The fluid samples are placed with a pipette and removed with cotton swabs.

It is to be noted, that in this technology the characteristic impedance Z_0 in assumed to be purely real and a value of $50\,\Omega$ is used. The produced structures can have a different Z_0, e.g., $45\,\Omega$, but this is only a normalization factor and does not influence the accuracy of the extracted permittivity. It is also to be noted, that the simulated and measured sensors are to be calibrated with their respective calibration curves due to manufacturing tolerances.

3.4.1 First Version up to 12 GHz

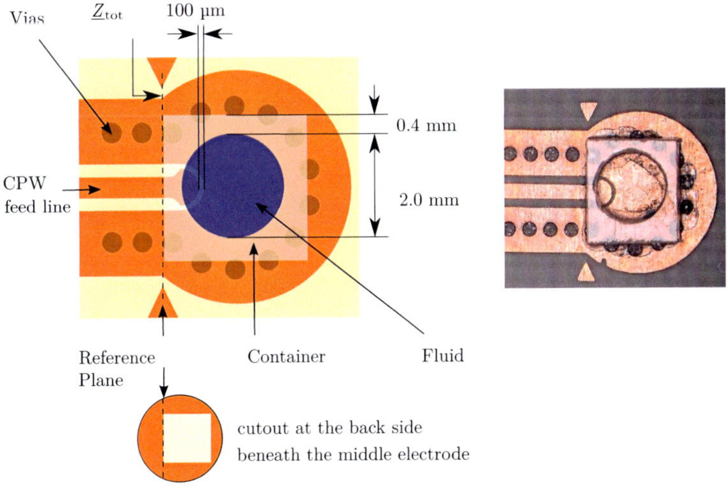

Figure 3.6: First PCB sensor with an operating frequency range up to 12 GHz.

The first sensor is presented in [68] and has a upper frequency limit of 12 GHz. In this version, the parasitic inductance is determined in a purely simulative way by using the inductance per unit length L' of the transmission line.

Sensor Design

The sensor design is based on a single transmission line and a single reflection at the MUT. It is basically a PCB counterpart to the OEC, but for smaller fluid samples.

The sensor is depicted in Figure 3.6. It is realized on *Rogers 4003* substrate, with an acrylic glass container glued on top. The feed is a CPW with an additional ground plane at the bottom. The connection to the network analyzer is ensured via microwave probes. To implement the simple equivalent circuit representation, the reference plane is situated as close to the MUT as possible. A thickness of 0.4 mm is achieved for the container walls. The sensor size behind the reference plane is about 0.9 mm x 0.9 mm large. The glue is positioned with a fine-placer to minimize spillover. To maximize the sensitivity, the distance between the middle electrode and the opposing ground plane is 100 µm, which is about the minimal achievable distance in this production process. The copper is 35 µm thick. The structuring tolerances are in the order of 10-20 µm. The container is 2 mm high, which is much larger than the penetration depth of the electric field. Therefore, the results are independent to small filling volume variations. The cutout in the ground plane at the back side of the substrate, right beneath the middle electrode, reduces the parasitics. The via wall ensures that no unintentional parallel plate modes are excited. The sensor does not radiate nor resonate because it is electrically small. In principle, a CPW sensor without an additional ground plane functions as well, as demonstrated in [68].

Measurement results

As extracted from simulation, the parasitic inductance L_{par} is 285 pH. When neglecting it, the sensors operating frequency range is reduced from 12 GHz to about 3 GHz, as demonstrated in [68]. The other parasitics, namely C_{par} and G_{par}, vary with frequency and are extracted by using solely the measurement results of the reference materials. They include the contributions of substrate, glue, and acrylic glass because the electric field lines partially go through these materials. At 1 GHz the slope $\left(\frac{A}{l}\right)_{\mathrm{MUT}}$ is 1.06 mm, which will be used to compare the sensitivity of different sensors.

Figure 3.7 reports the permittivity values extracted from measurements. In the frequency range up to 12 GHz the deviations from literature values are smaller than 1 for both the real and the imaginary part. The saline solution with the highest concentration, namely 3 %, shows larger deviations. This is probably due to the high loss with $\tan\delta > 1$ which changes the electric field distribution more than fluids with low loss.

Simulation results, not shown here, exhibit a higher accuracy. This is to be expected because there are only two error sources, namely the simulation accuracy and the

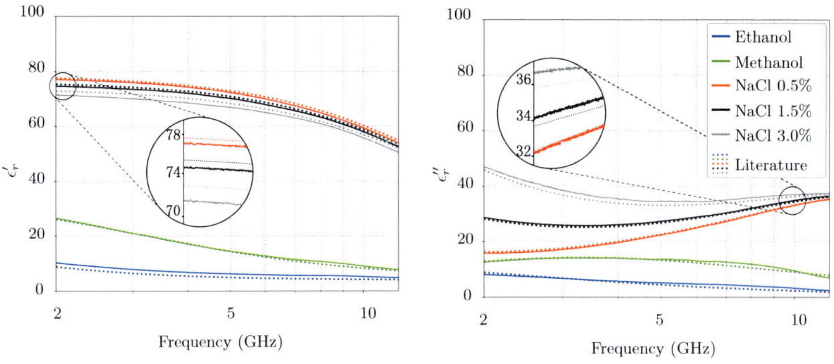

(a) Real part of the permittivity.

(b) Imaginary part of the permittivity.

Figure 3.7: Permittivity values extracted from measurements versus frequency for the first sensor.

model approximation. The permittivity of different MUTs is defined by the user and experiences no uncertainty. Deviations due to temperature fluctuations or general measurement inaccuracies are absent.

3.4.2 Second Version up to 20 GHz

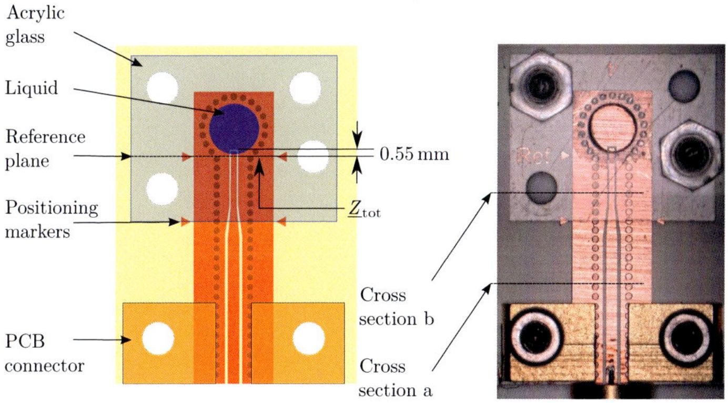

Figure 3.8: Second PCB sensor with an operating frequency range up to 20 GHz.

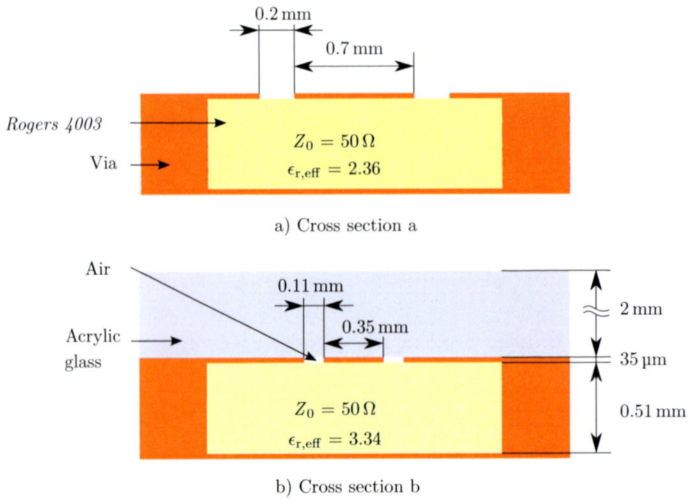

Figure 3.9: Different cross sections of the sensor feed line.

The sensing principle relies on the equivalent circuit representation, which is valid as long as the sensing area is electrically small. The basic assumption is that a decrease of the overall sensor size behind the reference plane increases the usable frequency range. To achieve this, the reference plane is placed closer to the electrodes and it intersects the acrylic glass container as depicted in Figure 3.8. The sensor is presented in [69] and has an upper frequency limit of 20 GHz. The choice to use a connector, as opposed to probes, is arbitrary. The use of screws allows for easier positioning of the container.

Sensor Design

This design requires a transition between the two cross sections which are sketched in Figure 3.9. Both are designed for a 50 Ω characteristic impedance but the cross sections differ due to a higher effective permittivity of the transmission line in the cross section b. The tapered transition has maximum reflections of -23 dB in simulation. The design allows to bring the reference plane closer to the MUT, resulting in a 0.55 mm x 0.57 mm area for the sensing part. To strengthen the mechanical stability, an additional 1 mm layer of an FR4 substrate is added below the bottom ground plane.

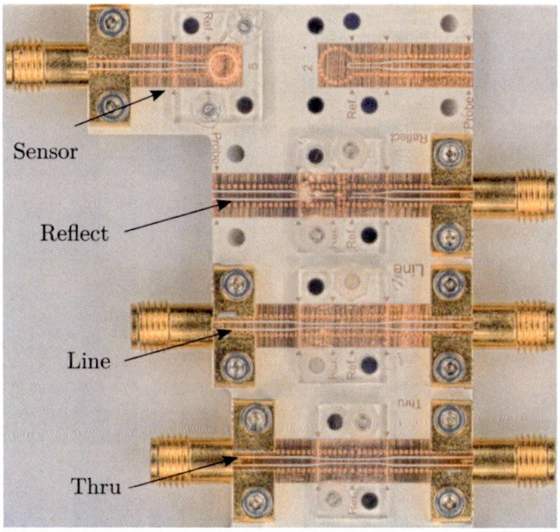

Figure 3.10: Second design of the sensor and the calibration kit.

Figure 3.10 shows a photography of the produced TRL calibration kit and the sensor. If more broadband measurements are required, additional line standards can be added for a multi-line TRL calibration [81].

Measurement results

A value of 125 pH for L_{par} is used in the following, as determined by simulation. The normalized total capacitance $C = C_{\mathrm{par}} + C_{\mathrm{MUT}}$ and conductance $G = G_{\mathrm{par}} + G_{\mathrm{MUT}}$ are plotted in Figure 3.11 versus the real and imaginary part of the permittivity, respectively. Hexagons filled with blue and red color, respectively, are the calibration standards. Other values are plotted to experimentally validate the predicted linearity. The slopes of the two curves, i.e., $\left(\frac{A}{l}\right)_{\mathrm{MUT}}$, are the same as assumed.

Due to the smaller size of the sensor compared to Section 3.4.1, the sensitivity $\left(\frac{A}{l}\right)_{\mathrm{MUT}}$ is reduced to 0.58 mm. The contrast between the total capacitance with air and with NaCl 1% as MUT is 0.4 pF. Figure 3.12 presents the experimental results. The reduction in size results in an increase of the upper frequency limit to around 20 GHz here. Notably, the design of the sensor is accomplished without knowing the electrical properties of the applied glue, which underlines the robustness of the calibration procedure. Above 15 GHz, some ripples in the extracted values can be seen. The reason for this effect is a possible inaccurate TRL calibration kit [82], [83]. Neverthe-

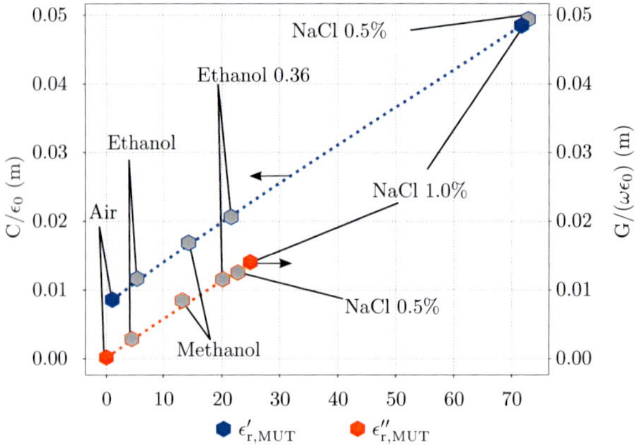

Figure 3.11: Total (normalized) C and G values extracted from measurements versus permittivity for the second sensor at 5 GHz.

less, improvements over the first sensor version are visible and further enhancements are made in Section 3.4.3.

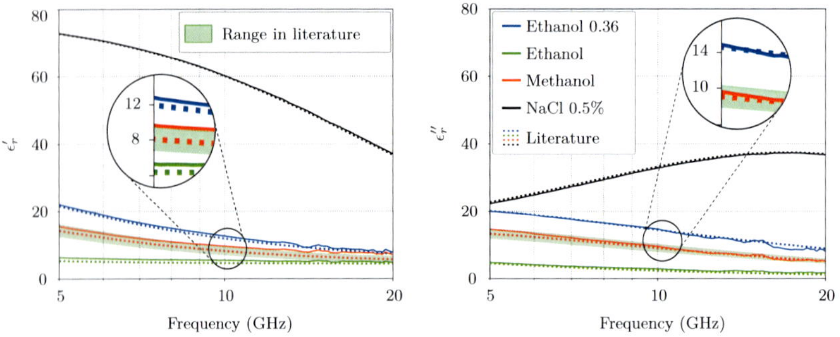

(a) Real part of the permittivity.

(b) Imaginary part of the permittivity.

Figure 3.12: Permittivity values extracted from measurements versus frequency for the second sensor.

3.4.3 Third Version up to 24 GHz

With increasing frequency, a long feed line can cause ripples in the measurements results [83]. A slightly modified version of the sensor from Section 3.4.2 is used to overcome this issue. The feed line between the reference plane and the connector measures 8.5 mm instead of 14 mm. Also, slightly wider CPW gaps, of 200 µm instead of 100 µm, combined with a thinner substrate are used to reduce the effect of production tolerances.

To further increase the frequency range, the parasitic inductance L_{par} needs to be known as precisely as possible. This is due to its impedance $\mathrm{j}\omega L_{\mathrm{par}}$, which increases its influence at higher frequencies. The method proposed in Section 3.2.2 is applied here.

Inductance determination

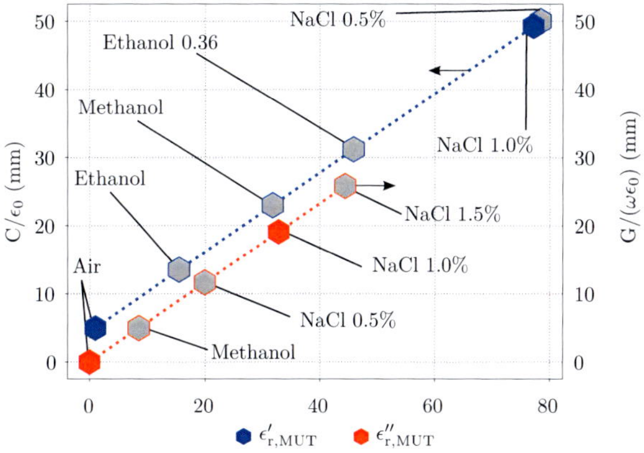

Figure 3.13: 1 GHz normalized simulation results for C and G of the PCB sensor for different liquids. The calibration standards are marked as blue and red filled hexagons. For the sake of clarity, not all materials are plotted for both C and G due to overlapping.

Simulation results are used to demonstrate the applicability of the proposed approach as they avoid experimental uncertainties. As shown in Figure 3.13 the calibration curves are almost perfectly linear with the same slope $\left(\frac{A}{l}\right)_{\mathrm{MUT}}$ for the normalized C and G at 1 GHz, even without the inclusion of L_{par}.

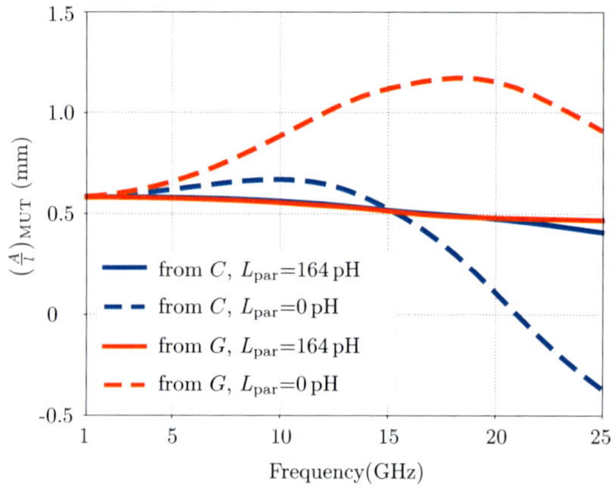

Figure 3.14: Extracted slopes $\left(\frac{A}{l}\right)_{\mathrm{MUT}}$ in simulation by using different values for the parasitic inductance.

Figure 3.14 reports the slopes over frequency for two different L_{par} values. With 0 pH, the slopes from C and G start to significantly deviate from each other at about 3 GHz. This yields an upper frequency limit of around 3 GHz as well. Using the simulated inductance per unit length, a value of 166 pH for L_{par} is found, which extends the operating range to 24 GHz.

Equation 3.10, evaluated from 1 GHz up to 24 GHz, yields 164 pH. This is close to the 166 pH estimation from the inductance per unit length. The performance of the sensor is demonstrated in Figure 3.15 with the permittivity extraction of methanol. By using L_{par}=0 pH, the deviation exceeds $\Delta\epsilon_{\mathrm{r}} = 1$ above 4 GHz for the imaginary part. With L_{par}=164 pH, the permittivity extraction is accurate with $\Delta\epsilon_{\mathrm{r}} < 1$ up to around 24 GHz for both the real and the imaginary part. This shows that the proposed method is applicable.

While the approach with Equation 3.10 does not offer a significant improvement for the simulated sensor, it does improve the practical performance. The simulated and the produced sensor are slightly different due to production tolerances and the reference plane is in practice not set as perfectly as in simulation. Thus, L_{par} values are different for the two cases. For the produced sensor a value of 143 pH is found with Equation 3.10, compared to 164 pH for the simulated sensor.

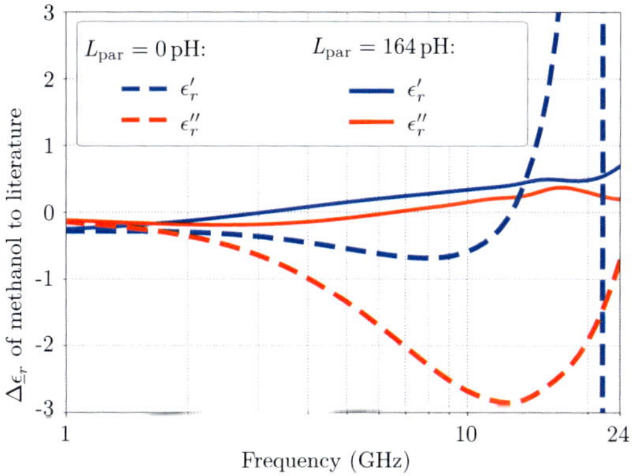

Figure 3.15: Difference between the extracted permittivity values of methanol in simulation and as reported in literature for different L_{par} values.

Measurement Results

The experimental results of the third sensor version are depicted in Figure 3.16 with L_{par}=143 pH. A multi-line TRL kit is used to cover the frequency range from 1 to 24 GHz. The applicable frequency range is increased without reducing the size of the active sensing part. The ripples reported in Section 3.4.2 are clearly removed. NaCl solutions show the best agreement with the literature values. Methanol values are within the ±1 range from literature for the most part and the ethanol permittivity is slightly overestimated.

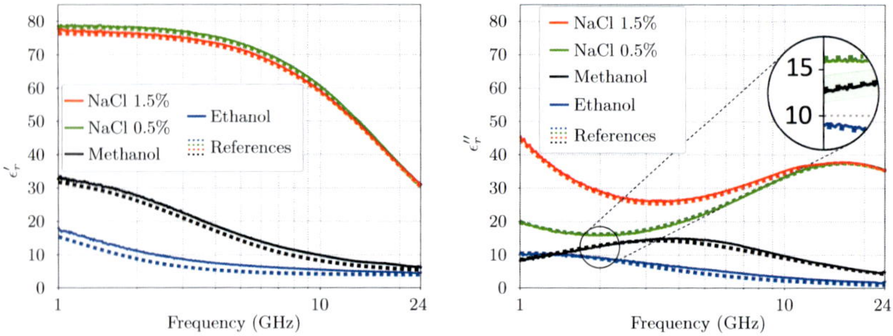

(a) Real part of the permittivity.

(b) Imaginary part of the permittivity.

Figure 3.16: Permittivity values extracted from measurements versus frequency for the third sensor.

3.5 Microchip Sensor

With microchips, measurements down to picoliter liquid volumes are possible. Microtechnology also allows for structuring resolutions in the order of several micrometers, enabling single cell measurements. It is essential to validate the applicability of the introduced two-step calibration procedure for this technology, especially for gold layers of less than 1 µm thickness.

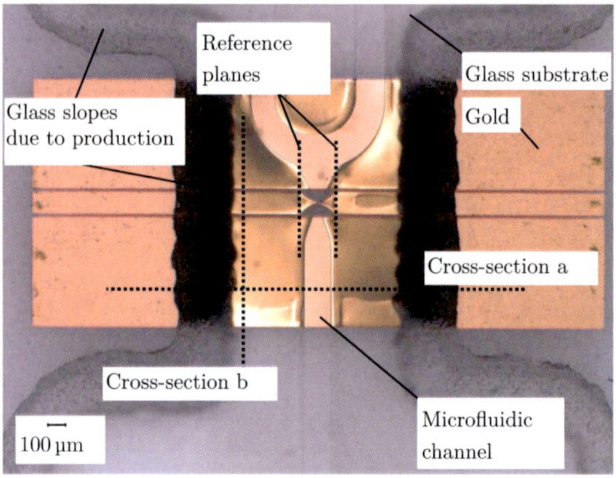

Figure 3.17: A top view photography of the analyzed microchip.

The microchip from Figure 3.17 is available for tests and is analyzed. It is produced by Micronit (Enschede, The Netherlands) with a commercial process. Initially intended for other medical applications, the microchip is used here for the verification of the proposed calibration procedure.

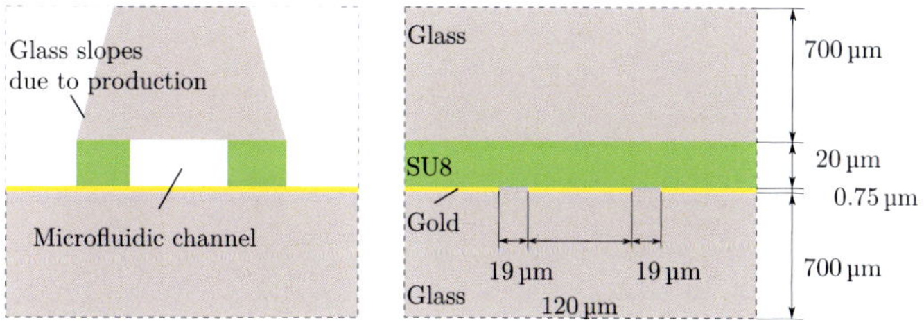

Figure 3.18: The two cross sections from Figure 3.17 (not to scale due to large size differences in horizontal and vertical direction).

The two marked cross sections are sketched in Figure 3.18, where the layer stack and the relevant dimensions can be seen. The 0.75 µm thick CPW leads to the MUT within the microfluidic channel. In this case, the MUT is the volume sample directly above the center electrode tips. The reference planes, which are 0.25 mm apart, are set with a multi-line TRL calibration kit.

For simulation purposes the permittivity of the glass is set to 3.2 with a $\tan\delta$ of 0.005, and that of the SU8 photoresist to 4.1 with a $\tan\delta$ of 0.015. The conductivity of the sputtered gold is around $3.5 \cdot 10^7$ S/m [55], which is slightly lower than for pure gold.

As this is a two-port scenario, the equivalent circuit is adapted as sketched in Figure 3.19. The resulting pi-circuit is easily extracted from the S-parameters. The elements $\underline{Y}_1$ and $\underline{Y}_2$, which are theoretically identical due to symmetry, are not needed for the permittivity extraction of the MUT. Otherwise, the procedure of extracting the permittivity from C_{MUT} and G_{MUT} is the same as for the 1-port equivalent circuit.

A TRL calibration kit for this sensor up to 32 GHz is used. The parasitic inductance is neglected here, because the smaller dimensions result in an increased $\frac{1}{\omega C}$ and decreased ωL [70]. For sensors with even higher frequency limits, the influence of the inductance needs to be re-evaluated, e.g., by means of simulation.

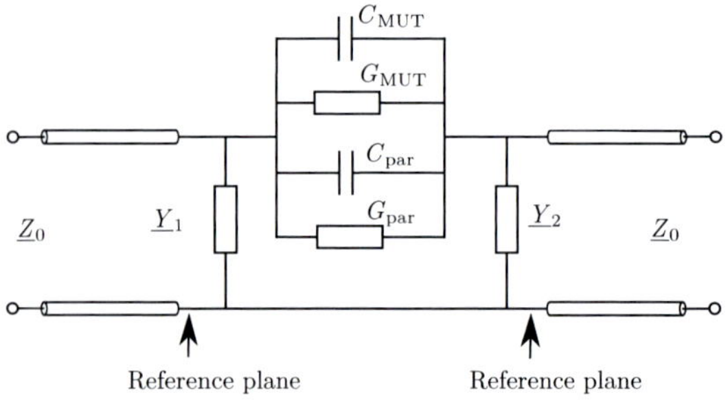

Figure 3.19: Equivalent circuit representation of the microchip.

3.5.1 Measurement Results

Because the metalization thickness is in the order of the skin depth, $\arg(\underline{Z}_0)$ needs to be known. It is extracted with the propagation constant $\underline{\gamma}$ as described in Section 3.2.3. Figure 3.20 reports three curves of $\arg(\underline{Z}_0)$ versus frequency up to 32 GHz.

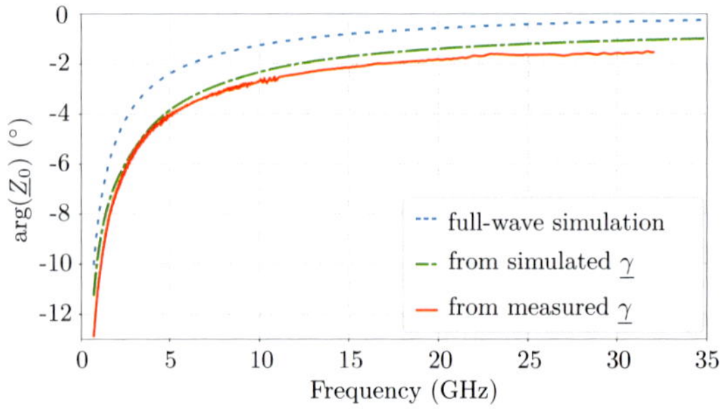

Figure 3.20: Simulated and extracted $\arg(\underline{Z}_0)$ from simulation and measurements.

The full wave simulation, represented with a blue dotted curve serves as reference. The values extracted with the simulated $\underline{\gamma}$ are represented by a green dashed curve. They show some deviations in the order of 1° from the full-wave simulation. This deviation is due to the neglected G'. It is shown in the following, that this accuracy is sufficient for the permittivity extraction. The values extracted with the measured $\underline{\gamma}$ are represented with a red curve. They are similar to the second curve. The remaining deviations can be due to uncertainties of the microwave dielectric properties of the microchip materials and to production tolerances.

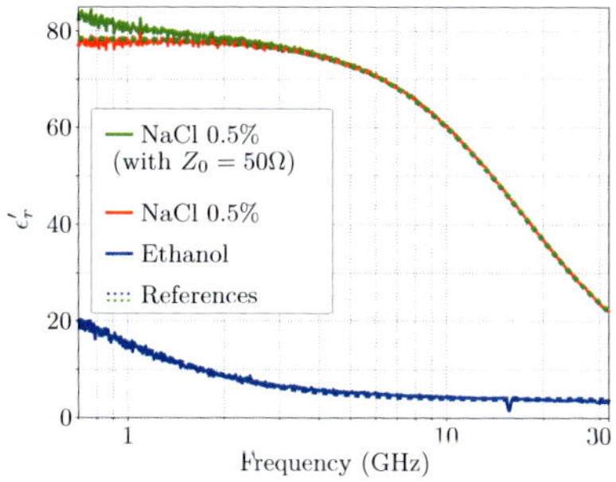

(a) Real part of the permittivity.

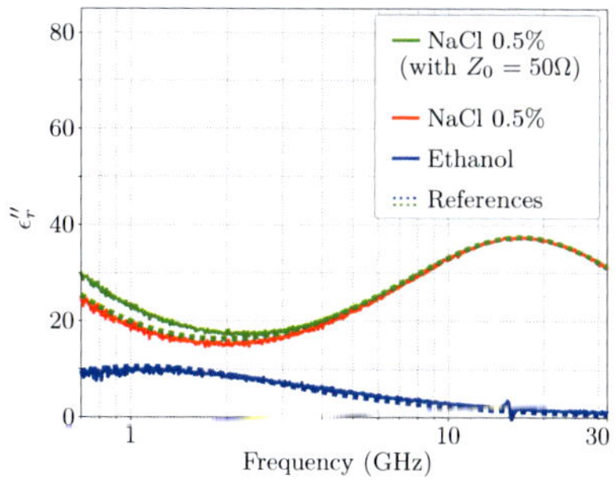

(b) Imaginary part of the permittivity.

Figure 3.21: Permittivity values extracted from measurements versus frequency for the microchip sensor.

Figure 3.21 reports measurements of two different materials, ethanol and NaCl 0.5%, with the microchip. The extracted permittivity is in good agreement with literature values from 0.7 GHz up to 32 GHz. A small ripple in the ethanol results at 14 GHz is an artifact of the multi-line TRL calibration. The capacitance difference between air and the NaCl 1% solution is around 5 fF and the slope of the calibration curves is approximately 7.1 µm. This is obviously sufficient for successful permittivity measurements at this size. It is expected that single cell sensors with similar electrode size show this kind of sensitivity as well. The integrated TRL calibration is limited to 32 GHz. However, the simulation indicates a potential operation up to 50 GHz for the sensor.

Additionally, the extracted permittivity of NaCl 0.5% with a real valued Z_0 is plotted in Figure 3.21 to emphasize the importance of the $\arg(\underline{Z}_0)$ extraction. The errors in the lower GHz range are obvious. The resulting deviations are plotted versus frequency in Figure 3.22. For the first several GHz, where $-\arg(\underline{Z}_0)$ is above 2°, they exceed $\Delta\epsilon_\mathrm{r} = 1$. Thus, the accuracy of the $\arg(\underline{Z}_0)$ extraction needs to be 2° or better for reliable permittivity results.

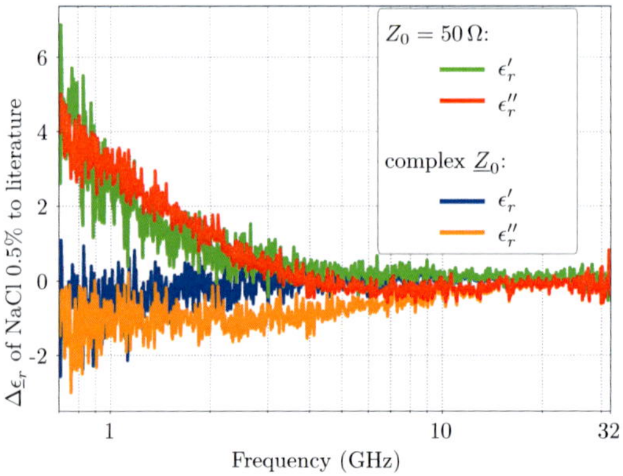

Figure 3.22: Deviations between the extracted permittivity values of a 0.5% NaCl solution to literature values.

3.5.2 Cole-Cole Parameters Fitting

To easily compare permittivity results with those obtained by other groups, Debye- or Cole-Cole-model parameters can be used. Additionally, such model parameters allow for simpler statistical analysis of repeated measurements. Here, the Levenberg-Marquardt algorithm [84], [85] is applied on a basic least-squares curve fitting problem to extract the Cole-Cole-parameters for a NaCl 0.5% solution. The measurement results are taken from the microchip experiments in Section 3.5. The parameters are presented in Table 3.1, together with literature values from [27]. The extracted parameter α equals zero here, which corresponds to the Debye-model. To improve the accuracy of ϵ_∞, more broadband measurements are necessary. With an upper frequency limit of 32 GHz, the presented value of ϵ_∞=6 is only an estimation. Other values, namely ϵ_{static}, τ,and σ_{DC}, are more similar to literature values. The resulting deviations are indications of the average error for various frequency regions, see Equation 2.2. Note that in [27] the parameters are obtained with measurements up to 20 GHz.

3.5.3 Uncertainty Analysis

Because the sensor and volume samples are small, the microchip is notably affected by measurement errors. In order to analyze the uncertainty, repeated measurements

Table 3.1: Cole-Cole-model parameters for a NaCl 0.5% solution.

Value	Values from [27]	This work
$\epsilon_{\mathbf{static}}$	78.66	77.99
ϵ_{∞}	5	6
τ	9.15 ps	9.17 ps
α	0.005	0
$\sigma_{\mathbf{DC}}$	0.87 S/m	0.82 S/m

are taken within one hour at room temperature. With an input power of 0 dBm, the 1% NaCl solution is measured thirteen times. During this measurement window, the fluid is flowing constantly and the measurement probes are not moved. These results are then compared to the initial 1% NaCl calibration measurement. Figure 3.23(a) shows the mean difference and standard deviation of the $\underline{S}_{12}$-parameter for both magnitude and phase of the thirteen 1% NaCl measurements. For comparison, Figure 3.23(b) shows the difference of $\underline{S}_{12}$-parameter between the two calibration standards, air and 1% NaCl. Most importantly, the resulting deviations in terms of the extracted permittivity are plotted in Figure 3.24. The deviations are mostly below 1, so the proposed procedure is applicable for repeated measurements with microchips within at least an hour. An uncertainty analysis with repeated measurements of a 0.5% NaCl solution has similar results and is not shown here.

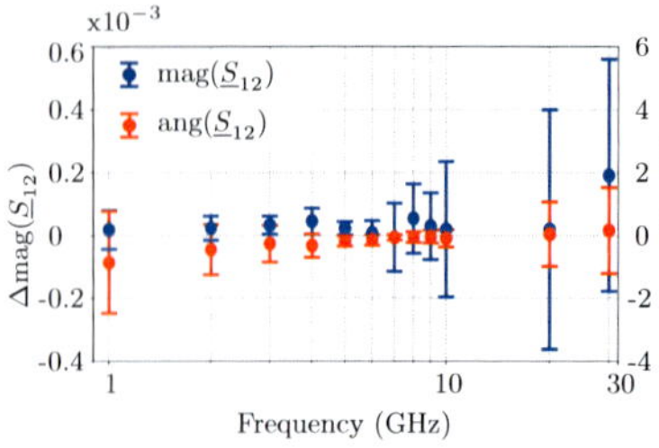

(a) Mean error and standard deviation for a 1% NaCl solution from 13 independent S_{12}-measurements compared to the initial calibration data.

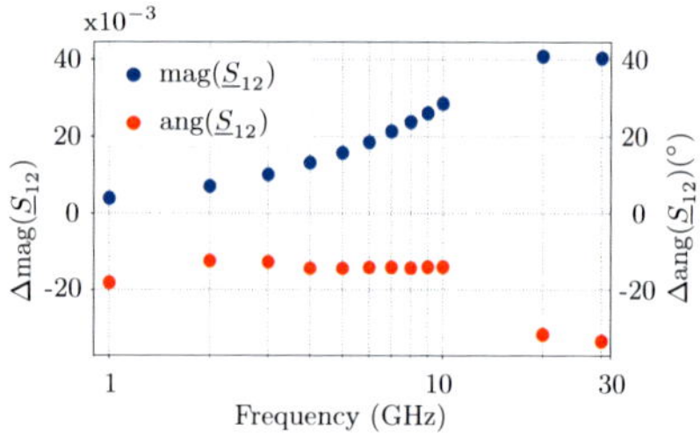

(b) Difference between the S_{12}-parameter of air and the 1% NaCl solution used for calibration.

Figure 3.23: S-parameter results for selected frequency points.

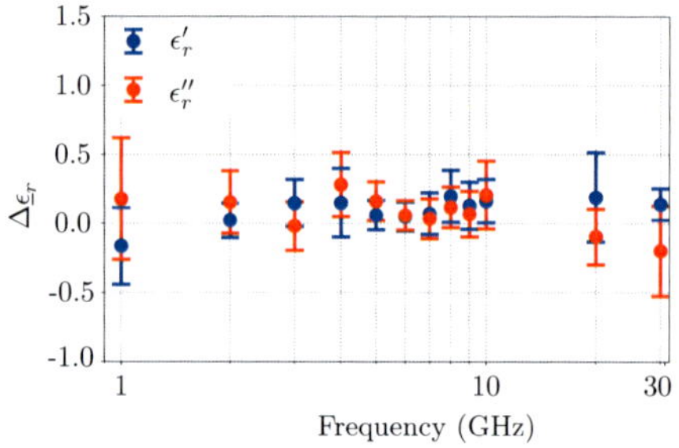

Figure 3.24: Mean error and standard deviation of the permittivity values extracted from thirteen measurements of a 1% NaCl solution.

4 Planar Sensors for Single Cell Measurements

In order to determine the complex permittivity of single cells, small electrode dimensions in the order of a single cell radius are required. To achieve this, microchip technology, as tested in Section 3.5, is used. Two sensors are developed, tested and compared in regard of their practicability and accuracy. One of them is a 1-port device, the other has two ports. Two cell lines with three different versions each are available for experiments. This ensures a large variety of measurements.

The basic concept is to accurately position the cells on top of the electrodes and utilize the calibration procedure from Chapter 3. For this to work, the electrodes are made smaller than the cell diameter. Otherwise the cell size would play a role and make the calibration needlessly difficult. Many options for mechanical cell trapping exist, for example [86] for vertical and [87] for horizontal positioning. Here, a three-way mechanical trap with horizontal and vertical positioning is designed for both ease of use and ease of production.

In Sections 4.1 and 4.2 the two planar sensor designs and their measurement setup are presented, respectively. Measurements of single CHO and HepG2 cells in different pathological states are carried out in Sections 4.3 and 4.4, respectively. Due to practical issues, like congestion, the number of measurements is not the same for every combination of sensor, cell type, and pathological state. An overview of all of the measured cell versions is presented in Section 4.5 in form of Cole-Cole parameters. Additionally, a comparison with literature values is given. The structure, design choices, measurement setup, and cell preparation are similar to [88].

All microchips in this chapter are produced by Micronit (Enschede, The Netherlands). All measurements are performed at room temperature, i.e., at 21°C.

4.1 Sensor Designs

In theory, any feed line can be used for the sensors. Because some experience is gathered with the use of CPW, the same is used in this chapter. The basic design idea stays the same – an electrically small sensing area represented by an equivalent circuit.

The microchips are produced on glass wafers with gold structures for the electrodes and a SU8 photoresist layer for the microfluidic channels. An exemplary cross-section is sketched in Figure 4.1. The materials involved are the same as in Section 3.5. The SU8 layer now has a thickness of 10 µm. Additional 10 µm are etched into the top glass substrate for a total of 20 µm, which is available for the microfluidic channel. This height is sufficient for single cells to move freely. Due to under-etching effects, the glass wall is rounded, whereas the SU8 wall portion is straight.

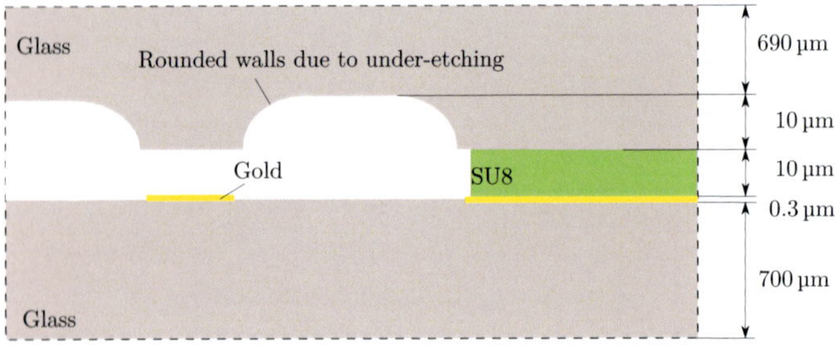

Figure 4.1: The principal structure layout available for the microchips.

With these design choices for the microfluidic channel, the cells can be squeezed vertically to ensure physical contact to the gold electrodes. The latter are now 0.3 µm thin, which allows for electrode widths and gaps down to about 3-5 µm. The positioning accuracy between the different layers is also about 3-5 µm.

1-Port Sensor

Figure 4.2 from the simulation with Ansys HFSS shows the 1-port sensor under different angles and emphasizes different layers. The sensing part consists of three electrodes which are fed via a CPW. The width of the gaps and the middle electrode in Figure 4.2(a) is 5 µm at the narrowest point. This ensures a physical contact of a squeezed cell and the electrodes. In Figure 4.2(b), the SU8 layer is narrowed down to 7 µm in

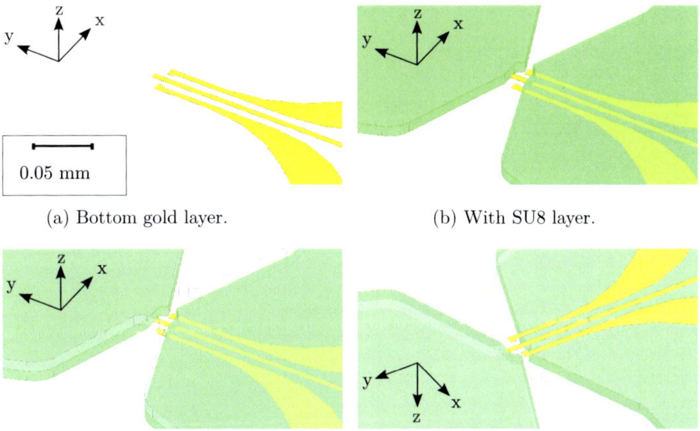

(a) Bottom gold layer.

(b) With SU8 layer.

(c) With top glass substrate.

(d) Flipped, with top glass substrate.

Figure 4.2: 1-port sensor design for single cell measurements.

the middle, so that cells are trapped on top of the electrodes. The top glass substrate is present in Figure 4.2(c), whereas Figure 4.2(d) depicts the chip from the bottom. The top glass layer is etched in the same positions as the SU8 layer to form the microfluidic channel, except directly above the electrodes. As described earlier, the cell can be trapped vertically this way. For the sake of clarity, the under-etching effect is not drawn in Figures 4.2(c) and 4.2(d).

The manufactured microchip is seen in Figure 4.3(a). It is 20 mm x 5 mm large. Two tubing holes are on the left for in- and output of fluids. The tubing hole on the right is for the trapping channel. The CPW is located in the middle. The middle line of the CPW is 120 µm wide and the gaps to the ground plane are 20 µm.

Figure 4.3(b) shows the microfluidic channel and the CPW in more detail. The reference plane is located approximately 0.4 mm before the sensing area. Figure 4.3(c) presents a zoomed in view on the electrodes.

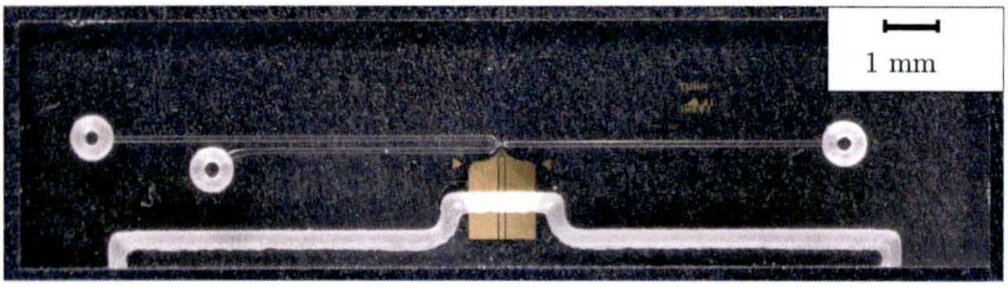

(a) Top view of the microchip.

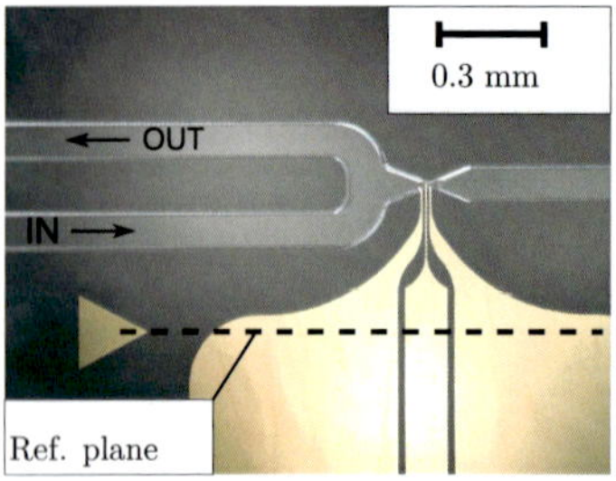

(b) Top view 1

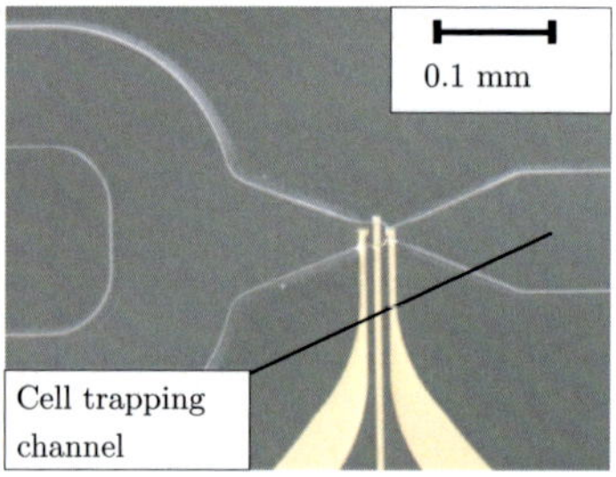

(c) Top view 2

Figure 4.3: Produced 20 mm x 5 mm 1-port microchip sensor.

2-Port Sensor

In analogy to Figure 4.2, Figure 4.4 shows the 2-port design with different layers. The feed CPW has the same dimensions. The middle lines of the CPW are narrowed down to form measuring electrodes as depicted in Figure 4.4(a). The two opposing electrodes have a width of 7 µm and a distance of 5 µm. The SU8 channel design in Figure 4.4(b) and the top glass substrate in Figure 4.4(c) are the same as for the 1-port version. Figure 4.4(d), again, shows the sensor from the bottom.

Similar 2-port sensors and electrode designs are published in literature [61], [62] with a different trapping mechanism. Another difference compared to this work is the calibration technique and the absence of broadband permittivity data.

A photography of the manufactured microchip is in Figure 4.5(a). The reference planes in Figure 4.5(b) are 0.5 mm apart. Figure 4.5(c) shows the electrodes as seen from the bottom.

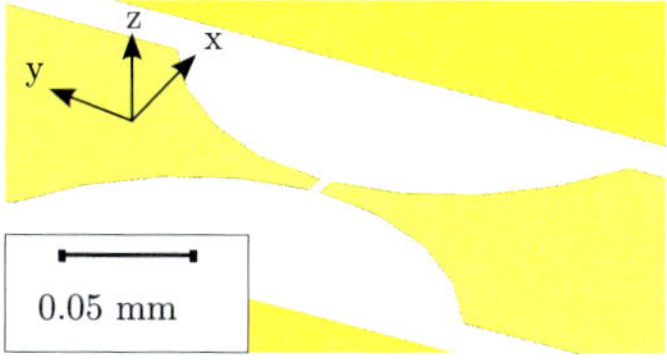

(a) Bottom gold layer.

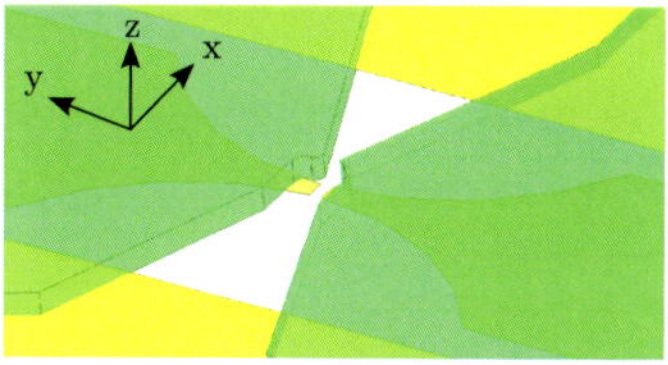

(b) With SU8 layer.

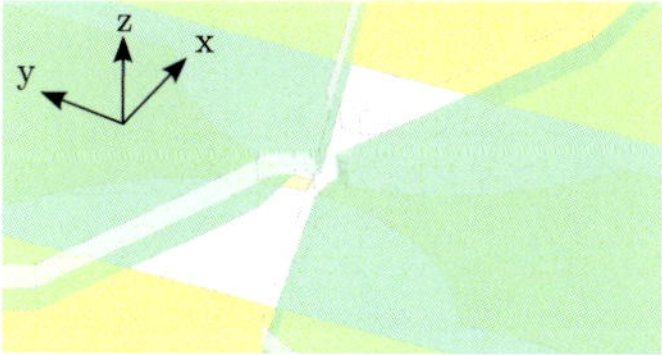

(c) With top glass substrate.

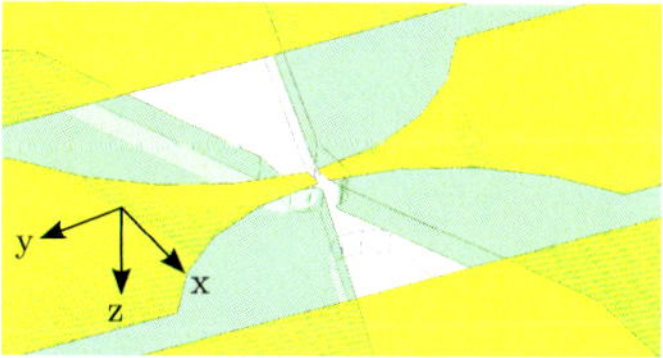

(d) Flipped, with top glass substrate.

Figure 4.4: 2-port sensor design for single cell measurements.

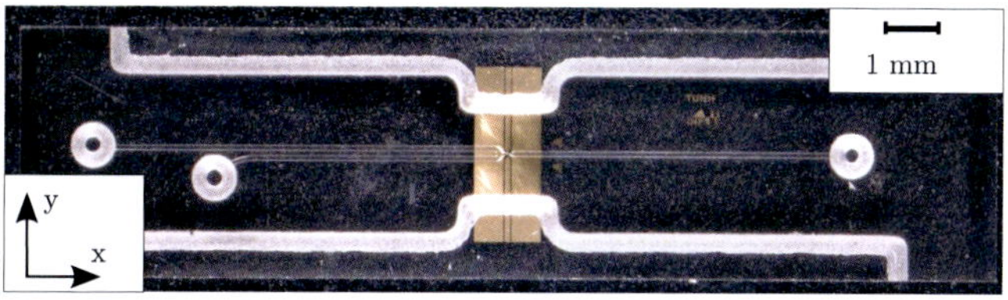

(a) Top view of the microchip.

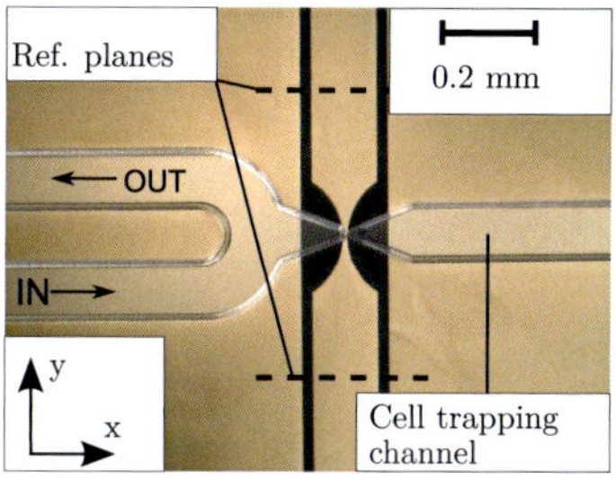

(b) Top view

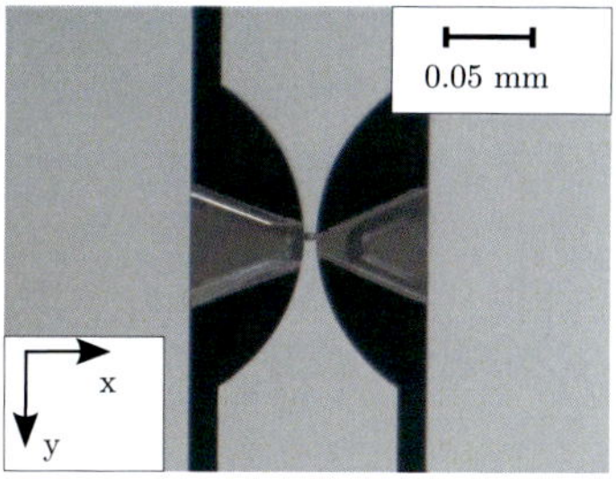

(c) Bottom view

Figure 4.5: Produced 20 mm x 5 mm 2-port microchip sensor.

4.2 Measurement Setup

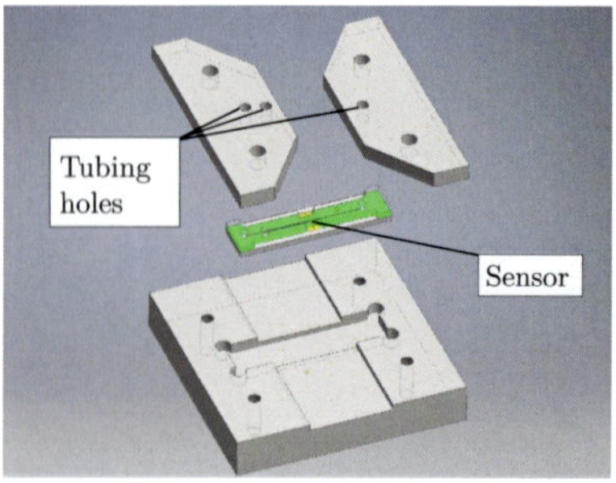

(a)

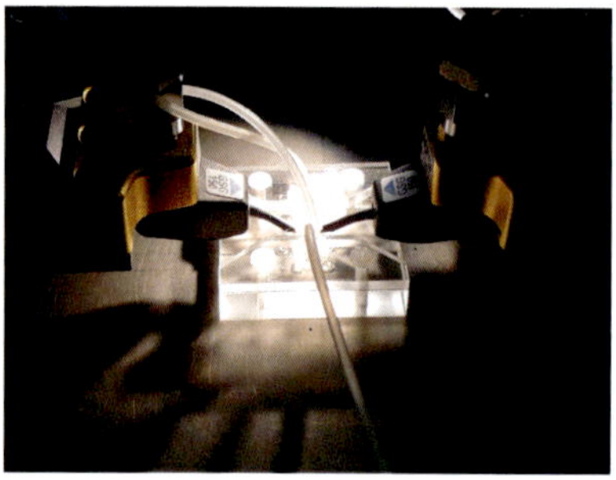

(b)

Figure 4.6: (a) Microchip with the chipholder and tubing attachments and (b) measurement setup with microwave probes attached.

The chip holder is made of plexiglass and is depicted in Figure 4.6(a) as drawn in Autodesk Inventor. It contains a cut out part in the middle for the microchip and holes to fasten the tubing connections. The tubes (not drawn) are attached to two plexiglass parts on the top with a bio-compatible silicone glue. Figure 4.6(b) shows an assembled measurement setup, including the tubing, with two probes connecting the microchip.

Simulation and Measurement of arg($\underline{Z}_0$)

Due to a thinner gold layer compared to Section 3.5, arg($\underline{Z}_0$) has to be re-determined. Figure 4.7 depicts the simulated and measured curves in the same manner as in Figure 3.20. The blue curve represents the exact values from the simulated cross-section. The green and red curves represent the values from a simulated and a measured line segment, respectively, extracted with the S-parameters and the propagation constant. The difference between the curves is less than 2° over the measured bandwidth, which was set as an accuracy goal in Section 3.5. The measurements of fluids and their permittivity determination with these sensors are similar to Section 3.5 and [70]. Thus, only cell measurements are shown in the following.

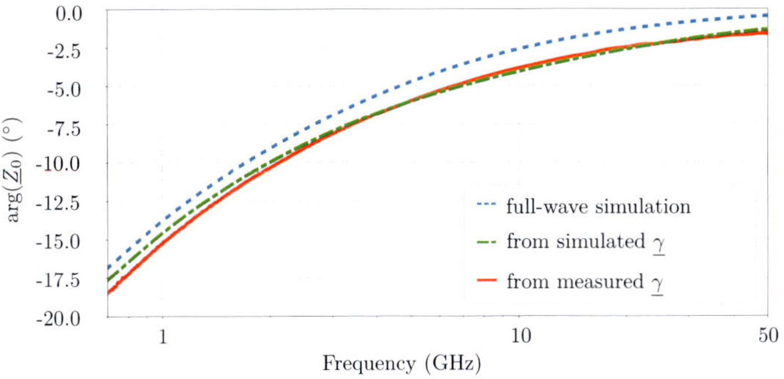

Figure 4.7: Curves for $\arg(\underline{Z}_0)$ from simulation and measurements.

4.3 CHO Measurements

Experimental Protocol

The experimental protocol is similar to [88] and is shortly described in the following.

The CHO DP-12 cells, from an IL-8-antibody-producing cell line, are cultivated in a TC-42 medium (Xell AG, Bielefeld, Germany), supplemented with 200 nM methotrexate and 8 mM L-glutamine in shaking flasks of 125 mL. The conditions for cultivation are 37 °C, 200 rpm, and 5% CO_2.

Three cell versions are prepared for the measurements. For each of the versions, the cells were centrifuged at 300 x g for 5 minutes. Before measurements, the living cells are re-suspended in 5 mL PBS. For two versions of dead cells, they are re-suspended in 5 mL 70% ethanol at -20 °C and a 4% formaldehyde solution (formalin), respectively. After repeated centrifugation at 300 x g, the dead cells are re-suspended in 5 mL PBS. Cooling on ice is provided before measurements.

Figure 4.8 shows a typical living CHO cell distribution based on their diameter. Most of the cells have a diameter of about 12 µm to 16 µm. Besides whole cells, many cell scrap particles are present and counted as well. This explains the increasing particle count below 10 µm. For dead cells, the share of scrap particles is larger and that of whole cells lower. The size distribution of dead cells is similar to living ones.

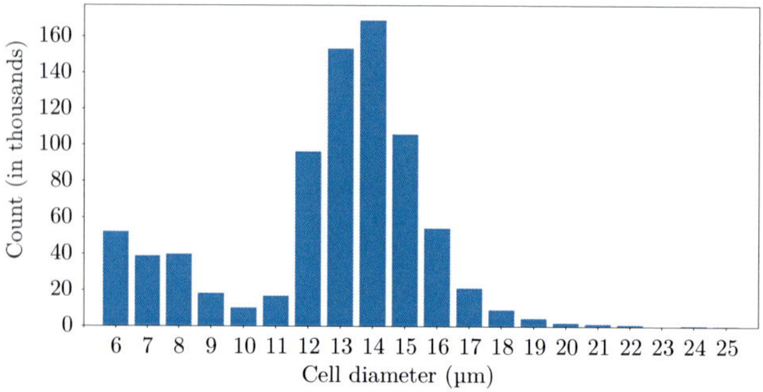

Figure 4.8: Living CHO cells cell count versus diameter.

1-Port Sensor

Figure 4.9 shows the trapping of a single cell via suction for the 1-port sensor. Due to the transparency of the cell, the visual contrasts are quite low. Therefore, the cell is marked with a dashed circle. In Figure 4.9(a) the cell is approaching the electrodes and in Figure 4.9(b) it is trapped. Practically, it often happens that the cells move fast and pass through the trap because of their elasticity or smaller than average size. In such cases, the cell can be pushed back by applying pressure with the trapping channel.

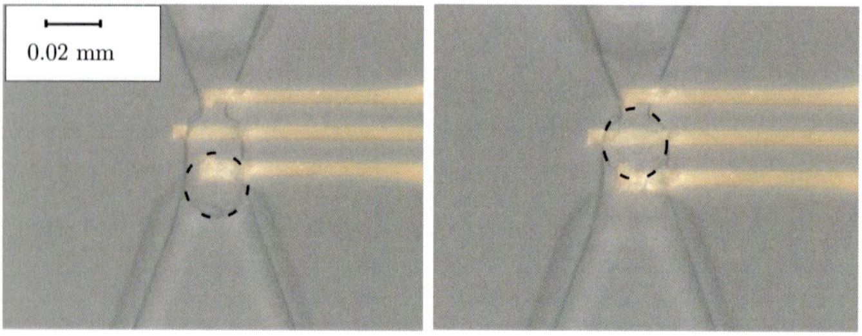

(a) (b)

Figure 4.9: Single cell trapping with the 1-port sensor with suction. The cell is approaching the sensing area (a) and is trapped right above the electrodes (b).

The permittivity results from single cell measurements are depicted in Figures 4.10(a) and 4.10(b) for the real and imaginary part, respectively. Green curves represent the results of two living cells. Red and black curves each contain the results of three

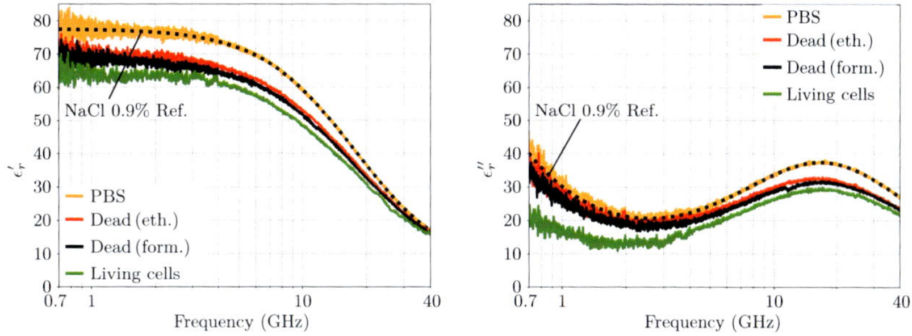

(a) Real part of the permittivity.

(b) Imaginary part of the permittivity.

Figure 4.10: Measured permittivity of living and dead CHO cells in PBS with the planar 1-port sensor.

dead cells affected by ethanol and formalin, respectively. Orange curves represent ten measurements of PBS. For comparison, the dashed line is the permittivity of the 0.9% NaCl solution used for calibration. The results exhibit a significant amount of noise in the lower GHz range, which decreases with increasing frequency. The permittivity values of PBS and the 0.9% NaCl solution are similar due to the same NaCl concentration in both.

Overall, the living cells show the lowest permittivity for both real and imaginary parts. They can be distinguished from other measurements at almost all frequency points due to the high contrast.

Dead cells show a lower, but still significant contrast to PBS compared to living cells at almost all measured frequency points. Additionally, the two versions of dead cells show a contrast between each other. Dead cells affected by ethanol have permittivity values closer to PBS compared to cells affected by formalin. Physically, ethanol affects and dissolves the cell membrane, whereas formalin dissolves the proteins. For this reason, is possible that cells after treatment with ethanol experience more material exchange with the PBS, which would explain the permittivity difference.

Figure 4.11 depicts the average contrast between the cells and the PBS as well as the resulting standard deviation. Below 5 GHz, the standard deviation is quite large and reduces with increasing frequency.

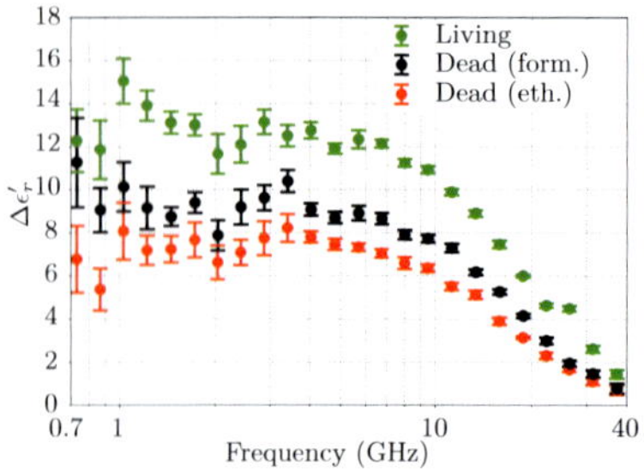

(a) Real part of the permittivity.

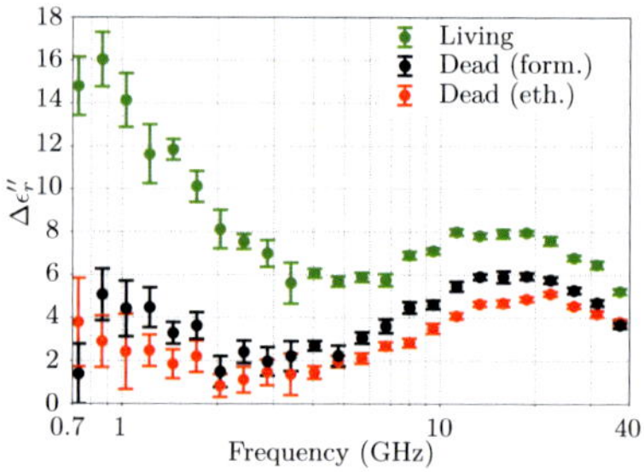

(b) Imaginary part of the permittivity.

Figure 4.11: Average difference between PBS and CHO cell permittivity results, alongside with the resulting standard deviation for the 1-port sensor.

2-Port Sensor

Figure 4.12 presents photographs of the cell trapping process with the 2-port sensor. They show a zoomed view on the measuring electrodes. A single cell is being moved towards the sensing area from the bottom in Figures 4.12(a) and 4.12(b). It is marked with a white dashed circle in Figures 4.12(c) and 4.12(d). Just like in the 1-port version, the cell is trapped and positioned on top of the electrodes by means of suction. The microscope settings here are different than in Figure 4.9 in order to better see the cell above the gold layer in Figure 4.12(a). As a result, the background appears black.

The measured real and imaginary parts of permittivity for the same three versions of CHO cells and PBS are plotted in Figure 4.13. Three cells of each type, along with ten PBS curves, are measured and depicted. The permittivity values are similar to the 1-port measurements, but contain less noise in the lower frequency spectrum.

In order to take a close look into the contrasts between the measurements, the average differences and the resulting standard deviations are depicted in Figure 4.14. Therefor, the permittivity of the cells is subtracted from the permittivity of the surrounding medium, i.e., the PBS, for many frequency points. For the real part of the permittivity, living cells reach a difference to PBS of about 15, dead cells of about 11 when affected by formalin, and of about 9 when affected by ethanol. In all cases, the difference to PBS decreases down to 2 at 40 GHz. For the imaginary part, living cells reach a difference of about 15 below 1 GHz, and of about 9 around 20 GHz. Dead cells show differences

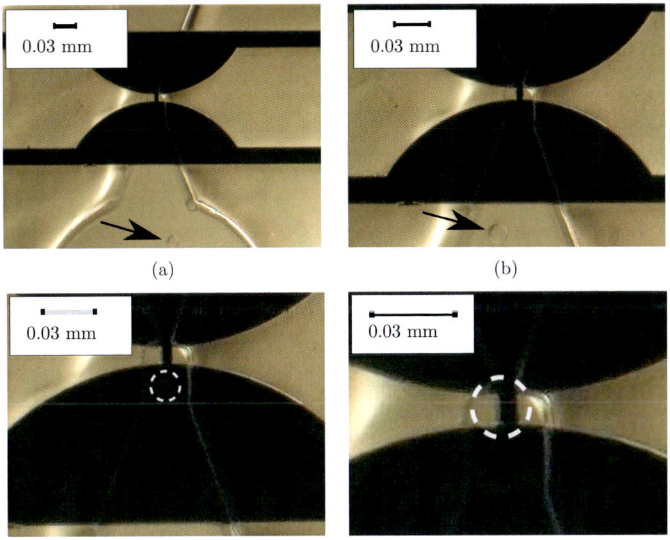

(a) (b) (c) (d)

Figure 4.12: Trapping a single cell with the 2-port sensor.

of about 7 and 5 below 1 GHz when affected by formalin and ethanol, respectively. Differences of about 6 are present at 20 GHz.

All in all, the measured permittivity contrasts of single cells to PBS are significantly higher than the typical accuracy of about one of such sensors [70]. The results even allow distinction between the different pathological states.

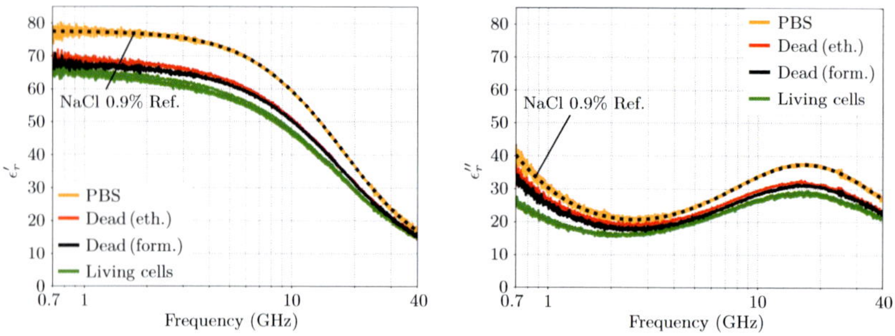

(a) Real part of the permittivity.

(b) Imaginary part of the permittivity.

Figure 4.13: Measured permittivity of living and dead CHO cells in PBS with the planar 2-port sensor.

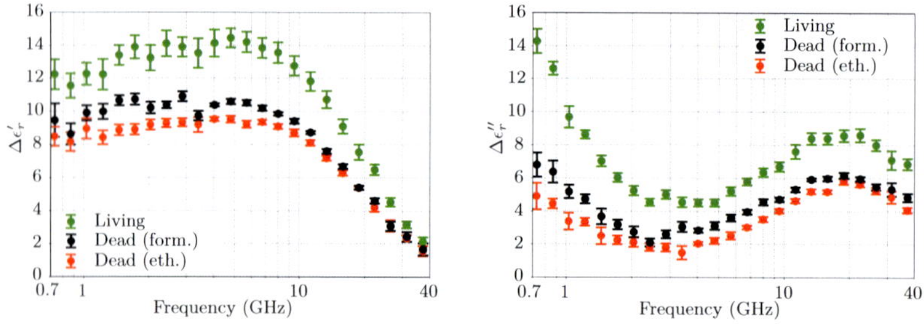

(a) Real part of the permittivity.

(b) Imaginary part of the permittivity.

Figure 4.14: Average difference between PBS and CHO cell permittivity results, alongside with the resulting standard deviation for the 2-port sensor.

p-Value Testing

A way to quantify the differentiability of the results is a t-test. The involved null-hypothesis in this case is that there are no differences between single cell and PBS measurements. The resulting p-values are calculated in the same way as in [88]. For chosen frequency points, the calculated p-values are presented in Table 4.1. Typical thresholds for significant results, i.e., rejection of the null-hypothesis, are 0.01 and 0.05 [89], representing 1% and 5%, respectively. All of the calculated p-values are below 0.01. Thus, the null-hypothesis is rejected.

Table 4.1: p-values for single cell measurements with the 2-port sensor. Values for the 1-port sensor are of similar magnitude.

Measurements	Frequency	p-value for ϵ_r'	p-value for ϵ_r''
Living cells	1 GHz	$8.1 \cdot 10^{-12}$	$8.3 \cdot 10^{-12}$
	15 GHz	$1.9 \cdot 10^{-14}$	$2.6 \cdot 10^{-14}$
Dead cells (formalin)	1 GHz	$6.2 \cdot 10^{-12}$	$4.4 \cdot 10^{-8}$
	15 GHz	$9 \cdot 10^{-15}$	$4.6 \cdot 10^{-14}$
Dead cells (ethanol)	1 CHz	$4.4\ 10^{-11}$	$2.5 \cdot 10^{-6}$
	15 GHz	$7.8 \cdot 10^{-15}$	$2.8 \cdot 10^{-13}$

4.4 HepG2 Measurements

Experimental Protocol

The HepG2 cells, a human liver cancer cell line, are aquired from the Leibniz Institute DSMZ, Braunschweig, Germany. They are expanded at 37 °C and a 5% CO_2 inadherent culture with RPMI culture medium (Biozym Scientific, Hessisch Oldendorf, Germany) supplemented with 10% fetal bovine serum (PAN Biotech, Aidenbach, Germany). Trypsin/EDTA (Biozym Scientific) is used for harvesting. With a density of 10^6 per mL and centrifugation at 300 x g, the cells are re-suspended in PBS. As in Section 4.3, the two versions of dead cells are re-suspended in 70% ethanol and a 4% formaldehyde solution (formalin), respectively. After a 5 minutes incubation, a washing with PBS buffer, the cells are re-suspended in PBS.

1-Port Sensor

Permittivity results for the HepG2 cells are depicted in Figure 4.15 for the 1-port sensor. Three living cells, three treated with formalin, and two with ethanol are measured a total of 15, 15, and 10 times, respectively. Additionally, ten PBS measurements are depicted. Living HepG2 cells show a larger contrast to PBS than living CHO cells. Otherwise, the results are similar to CHO measurements. Dead cells affected by ethanol

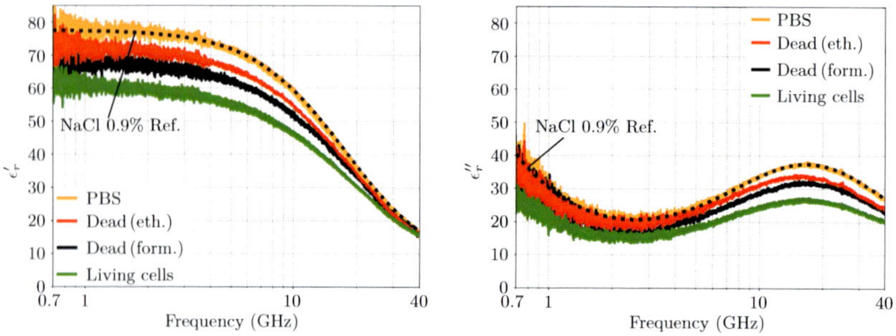

(a) Real part of the permittivity. (b) Imaginary part of the permittivity.

Figure 4.15: Measured permittivity of living and dead HepG2 cells in PBS with the planar 1-port sensor.

have less contrast to PBS than their formalin counterpart. Because the raw data is noisy in the lower GHz range here, a comparison between the two cells line is easier with Cole-Cole parameters as presented in Section 4.5.

2-Port Sensor

Figure 4.16 depicts the permittivity results for dead HepG2 cells as measured by the 2-port sensor. Four different cells treated with formalin and three with ethanol are measured. In total, the dead cell versions are represented with 18 curves each. For PBS, 30 curves are plotted. The permittivity values are comparable to those of the 1-port sensor and contain less noise at lower GHz frequencies.

A total of 40 measurements of six living cells is depicted in Figure 4.17. Two different patterns are formed. Because of overlapping values of one of the patterns with the curves of dead cells, the results of living cells are depicted in a separate Figure. One pattern, here named Group A, is similar to the measurements of living HepG2 cells with the 1-port sensor. The other one, Group B, has a significantly lower contrast to PBS and has values similar to dead cells. Not all cells survived the preparation process, thus the sample with living cells contains dead cells as well. In the following, Group A is assumed to contain the results of actually living cells.

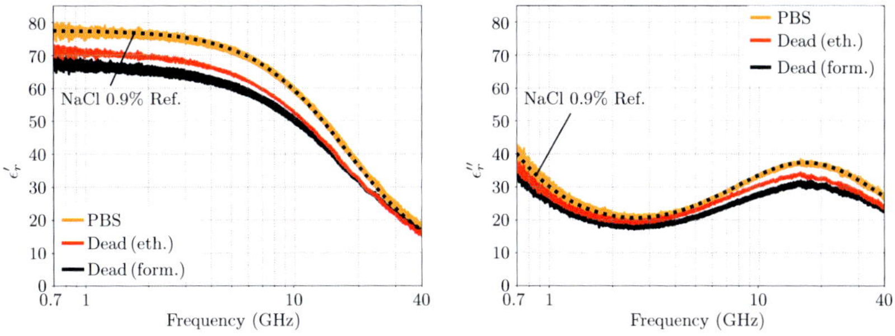

(a) Real part of the permittivity. (b) Imaginary part of the permittivity.

Figure 4.16: Measured permittivity of dead HepG2 cells in PBS with the planar 2-port sensor.

Average differences and the associated standard deviation between the PBS and HepG2 cells are depicted in Figure 4.18. Living cells reach a difference for the real part of the permittivity of about 21, which is higher than the measured 15 for CHO cells. In the lower GHz range, the two versions of dead cells can clearly be distinguished. After treatment with ethanol, the permittivity difference for the real part is about 7, whereas it is around 11 for formalin. Again, towards 40 GHz, the differences in the real part of the permittivity converge and the cells cannot be distinguished.

For the imaginary part, in Figure 4.18(b), the results for HepG2 cells are similar to CHO cells in the lower GHz range. Around 20 GHz, the imaginary part of the living cells reaches 12, whereas for CHO cells it only reaches 8.

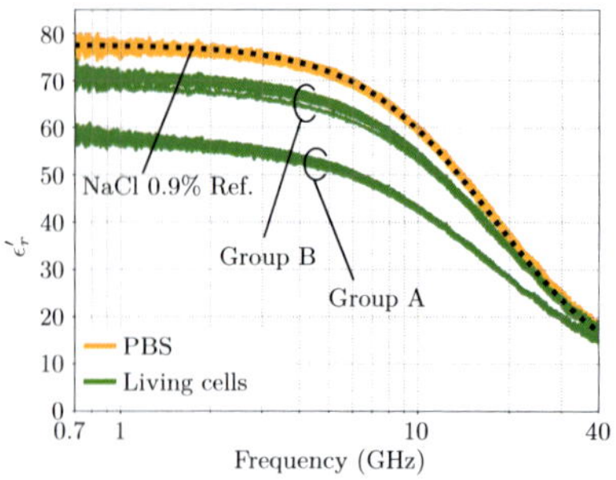

(a) Real part of the permittivity.

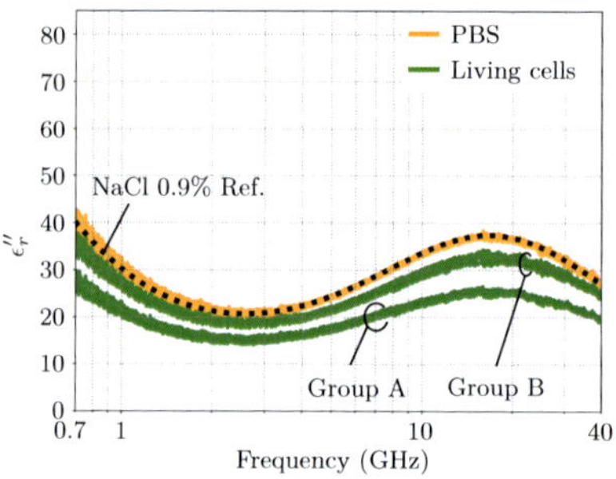

(b) Imaginary part of the permittivity.

Figure 4.17: Measured permittivity of presumably living HepG2 cells in PBS with the planar 2-port sensor.

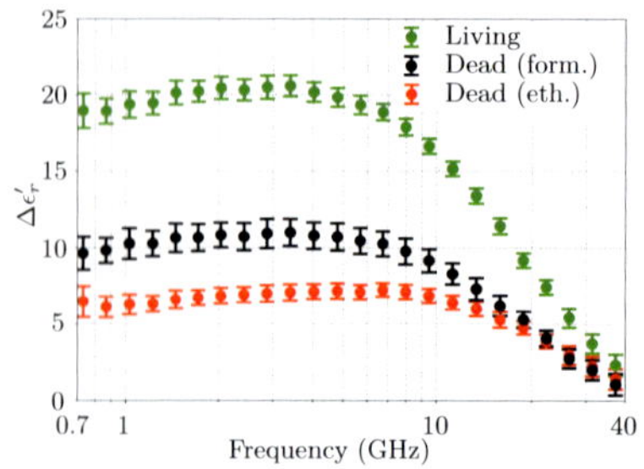

(a) Real part of the permittivity.

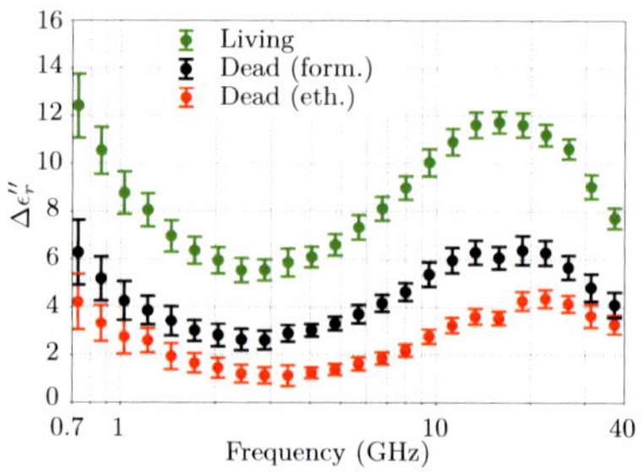

(b) Imaginary part of the permittivity.

Figure 4.18: Average difference between PBS and cell permittivity results, alongside with the resulting standard deviation for the 2-port sensor.

4.5 Cole-Cole Parameter Extraction and Discussion

For an easier comparison, the results from the 2-port sensor for CHO and HepG2 cells are provided in Figure 4.19 for living and in Figure 4.20 for dead cells.

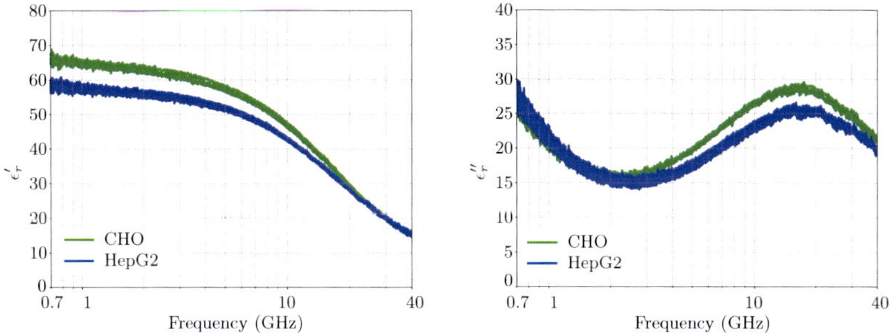

(a) Real part. (b) Imaginary part.

Figure 4.19: Measured permittivity of living CHO and HepG2 cells in PBS with the planar 2-port sensor.

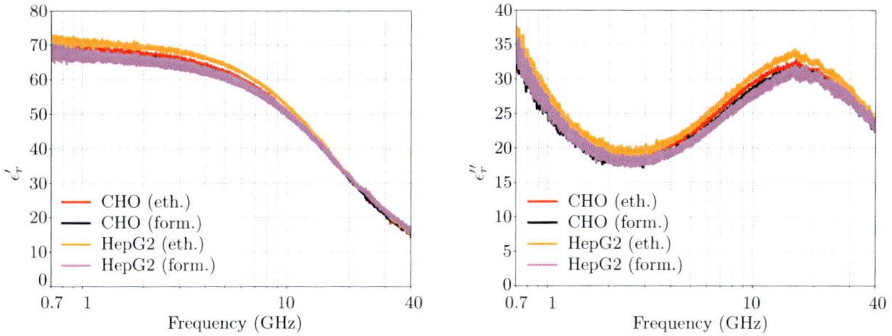

(a) Real part. (b) Imaginary part.

Figure 4.20: Measured permittivity of dead CHO and HepG2 cells in PBS with the planar 2-port sensor. The curves overlap for most frequencies.

The calculated Cole-Cole parameters for all measured combinations are presented in Table 4.2. They are only valid for the measured frequency range between 0.7 GHz and 40 GHz, which largely covers the γ-relaxation.

For the parameter ϵ_s, representing the upper limit of the measured ϵ_r curves, the results from the two sensor versions are in good agreement. Note that the accuracy of the measurement method is about $\Delta\epsilon_r \approx 1$ [70]. Living CHO cells have values of around 65, living HepG2 cells around 59. Thus, the two versions can be distinguished from each other with a contrast significantly larger than the measurement accuracy. Dead cells from both lines show values of about 67 to 68 when affected with formalin, and around 69 and 72 in the case of ethanol. Thus, the cell lines cannot be differentiated.

Table 4.2: Cole-Cole model parameters for cells at room temperature, valid between 0.7 GHz and 40 GHz.

	ϵ_s	ϵ_∞	τ	α	σ
1-Port, CHO:					
Living	65.59	4.49	9.35 ps	0.042	0.7 S/m
Dead (form.)	68.32	5.09	9.1 ps	0.031	1.27 S/m
Dead (eth.)	70.92	5.46	9.34 ps	0.033	1.32 S/m
2-Port, CHO:					
Living	64.6	4.06	9.6 ps	0.059	0.89 S/m
Dead (form.)	67.16	4.38	9.25 ps	0.03	1.2 S/m
Dead (eth.)	69.06	4	9.39 ps	0.042	1.26 S/m
1-Port, HepG2:					
Living	60.45	5.71	9.03 ps	0.04	0.97 S/m
Dead (form.)	67.96	5.3	9.02 ps	0.021	1.29 S/m
Dead (eth.)	72.22	6.15	9.46 ps	0.013	1.33 S/m
2-Port, HepG2:					
Living (only group A)	57.92	3.49	8.83 ps	0.075	0.98 S/m
(only group B)	70.84	4.46	8.81 ps	0.033	1.29 S/m
Dead (form.)	67.2	4.2	8.99 ps	0.046	1.22 S/m
Dead (eth.)	71.04	4.33	9.45 ps	0.034	1.3 S/m

The value ϵ_∞ shows results between around 3.5 and 6. Due to the frequency limit of 40 GHz, ϵ_∞ is basically extrapolated and thus not precise. The time constant τ shows little deviation between the cell variations. The exponent α is consistently higher for living cells compared to their dead counterparts. This means that the relaxation process for living cells is more stretched over frequency.

The conductivity γ for living cells is consistently lower than 1 S/m, whereas for dead cells it is higher than 1.2 S/m. For comparison, the conductivity of PBS is around 1.45 S/m.

Discussion

Only few microwave permittivity results are available in literature. In [2], CHO cells are measured using resonators at frequencies of 2.93 GHz and 5.73 GHz. For those two frequencies, the measured real part of the permittivity is around 78 and 71.5, respectively. This is significantly higher than 60 and 55, as measured here, respectively. The electrodes in [2] are wider than here, thus the measurements capture a significant portion of the surrounding medium. The resulting permittivity values are a mixture of the cell and medium permittivities. This prevents a direct comparison.

In case of CHO measurements in a cell culture, i.e., with many cells instead of a single one, a lower contrast to the surrounding medium is reported for dead cells compared to living cells [45], [54], [56]. This is the case in this work as well for single cell measurements. Significant dielectric contrasts between different cell lines are presented in [9] for single frequency points between 2 and 4.5 GHz.

Single cell microwave measurement results are available for THP1 cells, a human leukemia monocytic cell line. In [62], broadband capacitive and conductive contrasts are presented, both of which are sensor dependent and not comparable to permittivity results in this work. However, the cells are measured in two different states, living and after irreversible electroporation, which affects the cell membrane as does the treatment with ethanol here. The living cells showed a higher contrast to the surrounding medium compared to cells affected by electroporation. This is consistent with the observation in this work when comparing living and dead cells.

In [61], permittivity results at 5 GHz and 30 GHz for living and dead cells are presented. The latter result from treatment with saponin, which creates small holes in the cell membrane eventually leading to cell death. The values for the real part of the permittivity are higher than here. At 5 GHz, for instance, the ϵ_r' of the living THP1 cells is 63, whereas the living CHO cells measured here have values around 58 and living HepG2 cells have a value of 52. The imaginary part differs at 30 GHz compared to the cell lines measured here. Altogether, this underlines the differentiability between different cell lines.

Due to missing broadband results in literature, a comparison of other values, like α or σ is not possible.

Both sensors show consistent results and are suitable for single cell permittivity measurements. The 1-port design has some practical advantages compared to the 2-port version. These are a flexible electrode design, a smaller footprint, and easier handling due to only one probe required for connection. On the other hand, the 2-port sensor shows a significantly lower noise level.

Overall, the permittivity signature of single cells at microwave frequencies is sufficient to differentiate between the mentioned cell lines and between different pathological states as well.

Guard Electrode Concept in Microwave Measurements 5

The previously investigated planar sensors measure the permittivity of single biological cells from one side. The electric field strength declines towards the middle part of the cell. If the center of the cell is to be investigated with impedance measurements, the cell needs to be placed between two opposing electrodes which introduce electric field lines through the cell center. A major drawback of such a configuration is the contribution of fringing fields to the overall measured impedance. Thus, depending on the cell size, more or less of the culture medium is measured with those fringing fields. This makes the evaluation of the measured data and extraction of the permittivity of the cell difficult because the measurements depend on cell size.

To avoid this problem, a solution with a guard electrode is proposed. This method is used in the lower frequency range for the same purpose of avoiding the measurement of fringing fields when measuring liquids [90], [91] and has been known for decades [92]. In this section, the applicability of such guard electrodes for measurements at microwave frequencies is explored.

In order to better understand the design considerations and the functionality, the sensor development is described step by step in several sections. The basic principle for low frequencies and the high-frequency implementation are described in Section 5.1. Theoretical considerations based on simulation results of a single cell sensor are given in Section 5.2. A scaled PCB experiment is carried out in Section 5.3 to demonstrate the proof-of-principle. A microchip implementation is described in Section 5.4, along with the photographies of the designs and the measurement setup. Measurements of the two sensor versions are given in Sections 5.5 and 5.6.

This Chapter is based on the results published in [93] and [88]. As in the previous Chapter, all of the microchips are produced by Micronit (Enschede, The Netherlands).

5.1 Basic Principle

The basic principle of a guard electrode is first presented for low frequencies. Afterwards, the theoretical consideration and implementation ideas at high frequencies are presented.

Low Frequencies

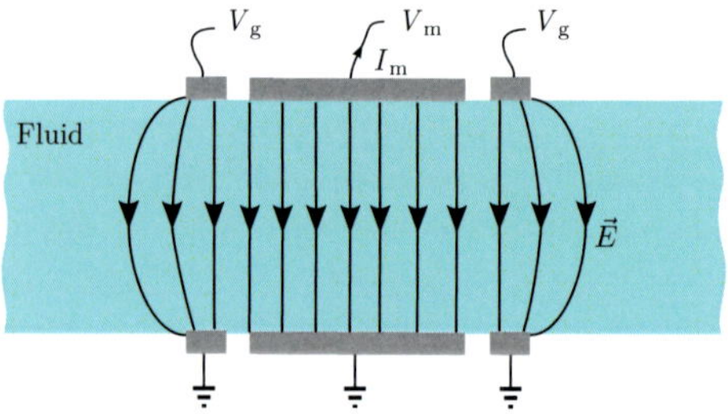

Figure 5.1: Exemplary low frequency implementation of a guard electrode with two synchronized voltage sources.

Figure 5.1 represents the basic principle of a guard electrode in a parallel plate capacitor configuration. The equal voltages V_m and V_g are delivered by two synchronized voltage sources. This way, the main and the guard electrodes have the same potential and no electric field lines between them. Due to the separate excitation, the capacitance for each of them can be evaluated independently. The field beneath the main electrode is essentially homogeneous, which leads to a simple parallel plate capacitance with the value $C = \epsilon_0 \cdot \epsilon_r \cdot \frac{A}{l}$ in the lossless case. Here, A represents the area of the electrode and l the distance to ground. The capacitance can be assessed with the admittance $\underline{Y}_m = j\omega C$ by evaluating V_m and I_m. For lossy materials the admittance $\underline{Y}_m$ has to be modified to include a conductance G to a more general expression

$$\underline{Y}_m = \omega\epsilon_0\epsilon_r'' \cdot \frac{A}{l} + j\omega\epsilon_0\epsilon_r' \cdot \frac{A}{l} = G + j\omega C. \tag{5.1}$$

High-Frequency Implementation

While voltages and currents are sufficient for a low frequency evaluation of a guard electrode scenario, high frequencies require the consideration of the incident and reflected waves. One possibility, which is explored here, is a two-port excitation. The equivalent representation is essentially a pi-circuit as depicted in Figure 5.2. The admittances of the main and the guard electrode are represented by $\underline{Y}_1$ and $\underline{Y}_2$, respectively. Their coupling is taken into account with $\underline{Y}_3$.

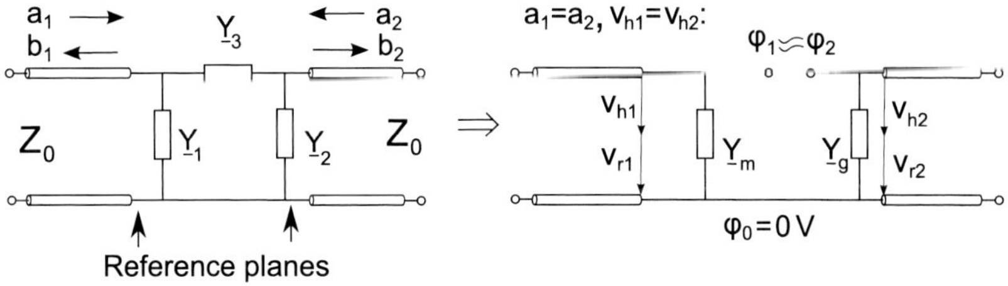

Figure 5.2: Exemplary equivalent circuit of a high-frequency implementation of a guard electrode. With equal excitation of both ports and a high impedance pi-circuit, the homogeneous and fringing fields are decoupled from each other.

In order to achieve equal potentials on both electrodes, the incident waves need to have the same power, i.e., $a_1 = a_2$. This results in equal incident voltages v_{h1} and v_{h2}. The reflection coefficients for both ports need to be approximately one, so that the reflected voltages are the same as well. Therefore, the impedances in the pi-circuit have to be significantly larger than the characteristic impedance Z_0 of the transmission lines. The incoming voltage is then effectively doubled on both sides. Thus, in the guard electrode configuration, the electrodes have the same potential with no field lines, nor current flowing between them. The remaining admittances are named $\underline{Y}_{\mathrm{m}}$, representing the admittance of the main electrode like in Equation 5.1, and $\underline{Y}_{\mathrm{g}}$, representing the admittance of the guard electrode.

The circuit is passive and linear. Therefore, superposition holds and the equivalent circuit elements can be calculated from the S-parameters. This simplifies measurements in practice because the ports do not need to be excited at the same time nor does their excitation need to be equal.

If only the MUT is measured with the main electrode, then its permittivity can be simply calculated from $\underline{Y}_\mathrm{m}$ from Equation 5.1. Due to the very small dimensions in the intended use case, the reference plane cannot be placed exactly in front of the biological cell. Inevitably, the main electrode will measure some parasitics, e.g., the substrate between the reference plane and the cell. The same principle as described in the calibration procedure in Chapter 3 is used to handle this problem. Therefor, the equivalent circuit is extended with the parasitic contribution $\underline{Y}_\mathrm{par}$ as depicted in Figure 5.3. The admittance $\underline{Y}_\mathrm{par}$, together with the contribution of the MUT, namely $\underline{Y}_\mathrm{MUT}$, represents the admittance of the main electrode. After calculating $\underline{Y}_\mathrm{m}$ with the scattering parameters, the process of extracting $\underline{Y}_\mathrm{MUT}$, and therefore $\underline{\epsilon}_\mathrm{r,MUT}$, is the same as in Chapter 3.

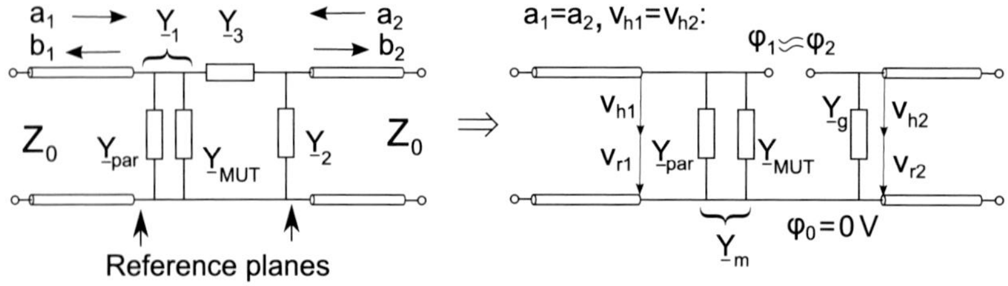

Figure 5.3: Modified equivalent circuit of a high-frequency implementation of a guard electrode. The main electrode admittance now consists of the contributions of MUT and parasitics.

5.2 Simulation Results

The theoretical considerations from Section 5.1 are first tested by means of simulation. Two sensors are designed, one with and the other without a guard electrode. Otherwise, their dimensions are exactly the same so that a comparison in regard of their permittivity extraction capabilities can be made. As test objects, spheres of approximately the size of biological cells are placed between the electrodes. A surrounding fluid with different permittivity definition than the sphere is added to evaluate the influence of fringing fields.

5.2.1 Sensor Without Guard Electrode

A possible parallel plate design without guard electrode as drawn in the simulation software Ansys HFSS is shown in Figure 5.4. It consists of a ground plane, a main electrode, and a biological cell squeezed in-between. The metal structures are manufactured on glass substrates, which are not shown here. An SU8 layer is used to manufacture microfluidic channels and single cell traps. For this design, the height of the SU8 layer should not be larger than the cell diameter. This way, the cell is squeezed, ensuring physical contact to the metal layers.

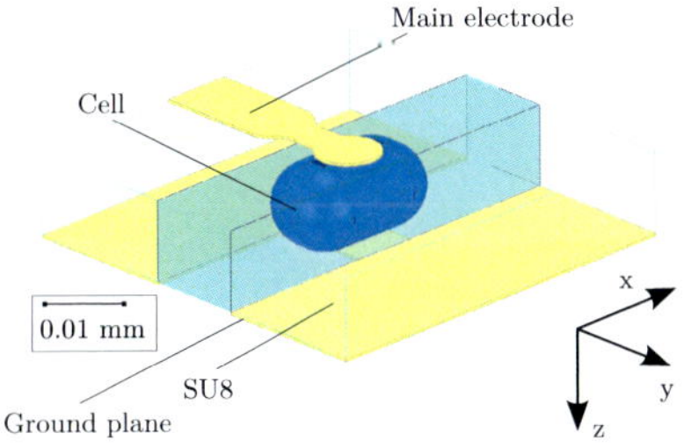

(a) Sensor design.

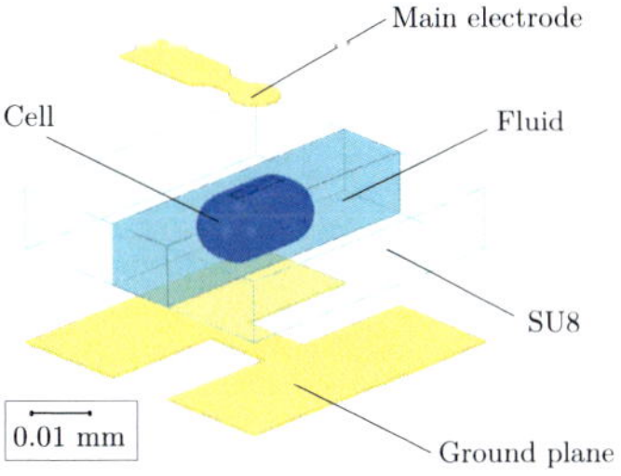

(b) Exploded view.

Figure 5.4: Exemplary single cell capacitive sensor in microtechnology.

Figure 5.5 shows the cross-section of the setup with the simulated electric fields. The electric field lines penetrate the middle of the cell as desired. Larger electrodes lead to higher capacitances, which increases the sensitivity of the sensor. However, some of the electric field lines penetrate the fluid around the cell. In practice, the cell size varies. Thus, the contribution ratio between the cell and fluid to the overall capacitance varies as well. The extracted permittivity of such a capacitance with mixed contribution does not represent only one material but a mixture of both as demonstrated in the following.

Figure 5.6 shows the permittivity extracted with the presented sensor from simulated S-parameters. The calibration method described in Chapter 3 is employed here. Air and a 0.9% NaCl solution serve as calibration materials and entirely fill the channel in the respective simulation. To demonstrate the sensor functionality, the channel is first completely filled with test fluids. Here, ethanol and NaCl 0.5% were used. After the simulation, the permittivities extracted from the S-parameters are well in agreement

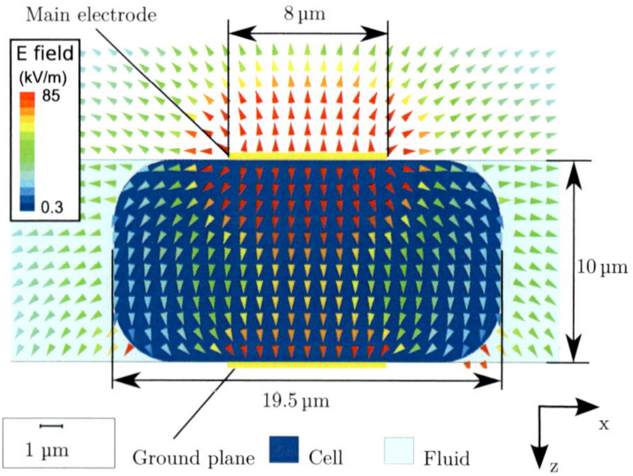

Figure 5.5: Electric field distribution in the cross-section of the single cell sensor without guard electrode.

with the material definitions. This shows that the sensor is in principle capable of accurately determining the complex permittivity of samples that completely fill the microfluidic channel.

For a single cell measurement scenario, the arrangement from Figures 5.4 and 5.5 is simulated. For demonstration purposes, the fluid is defined as ethanol and the cell as a NaCl 0.5% solution in simulation. This extreme contrast serves to show the influence of the fringing fields. In practice it might be lower, but the basic principle remains the same. In this case, as depicted in Figure 5.6, a mean permittivity value between those of the two materials is extracted. The result strongly deviates for both the real and the imaginary part of the permittivity. It is desired to solely measure the cell and extract its permittivity, which is set to NaCl 0.5% in this case. In practice, with different cell sizes, this deviation varies too.

One possibility to solve this issue is to decrease the distance between the electrodes. This reduces the contribution of the fringing fields but also means that the cell has less vertical space and is prone to get destroyed.

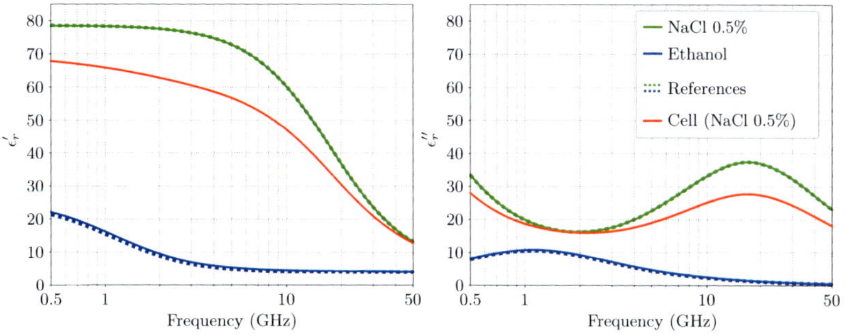

(a) Real part.

(b) Imaginary part.

Figure 5.6: Permittivity extracted from the simulated S-parameters for the single cell sensor. References are the material definitions in simulation.

5.2.2 Sensor With a Guard Electrode

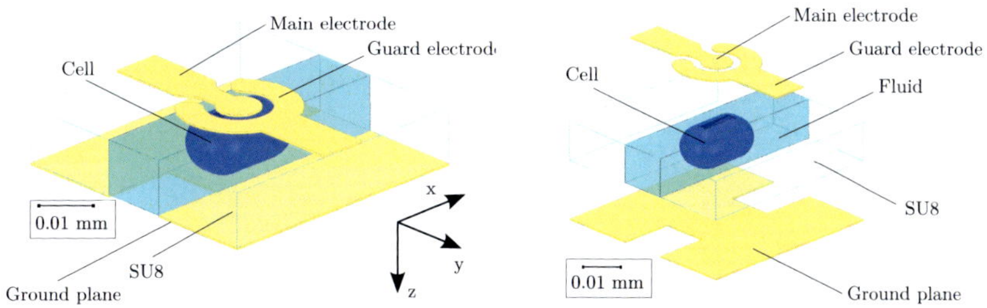

(a) Sensor design.

(b) Exploded view.

Figure 5.7: Exemplary single cell sensor with a guard electrode in microtechnology.

Figure 5.7 depicts the same sensor as in Section 5.2.1. but with an additional guard electrode. The latter is placed around the main electrode and is excited by its own port. Figures 5.8 and 5.9 illustrate the electric field distribution in two different cross-sections when both ports are excited with the same power. The middle of the cell exhibits an almost homogeneous distribution as desired. The fringing fields are successfully decoupled with the guard electrode. The equivalent circuit evaluation is done as described in Section 5.1. Otherwise, the calibration procedure remains the same.

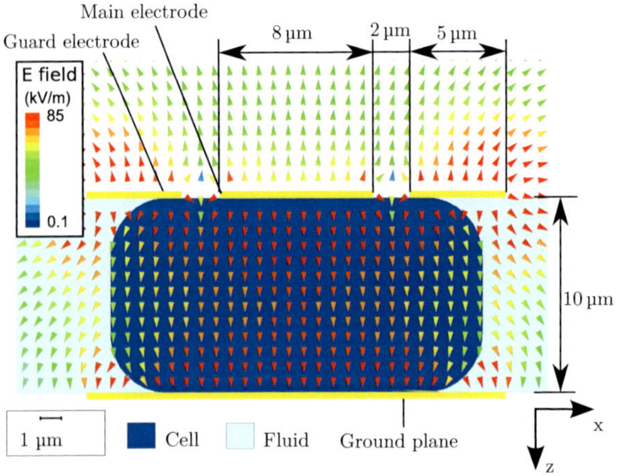

Figure 5.8: Electric field distribution in the x-z cross-section of the single cell sensor with a guard electrode.

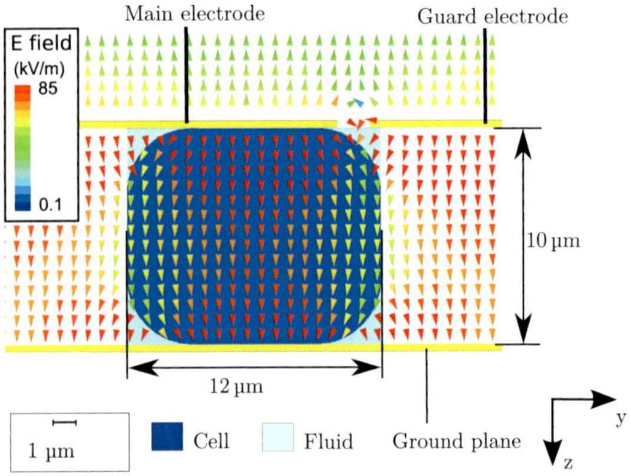

Figure 5.9: Electric field distribution in the y-z cross-section of the single cell sensor with a guard electrode.

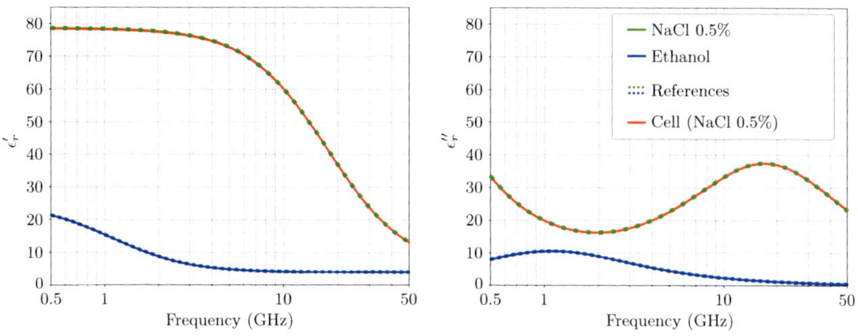

(a) Real part. (b) Imaginary part.

Figure 5.10: Permittivity extracted from the simulated S-parameters for the single cell sensor with a guard electrode.

The results of the permittivity extraction are depicted in Figure 5.10. When the channel is filled solely with a fluid, the results are as accurate as without the guard electrode. The values for the cell in a fluid coincide with those for NaCl 0.5%. Thus, the permittivity of solely the cell is correctly extracted, which demonstrates the functionality of the guard electrode in simulation.

5.3 PCB Experiment

For the experimental validation of the theoretical considerations of a guard electrode from Section 5.1, two scaled sensors in PCB technology are developed and tested. Similar to the approach from Section 5.2, two almost identical sensor are evaluated. Their only difference is that one has a guard electrode, the other does not.

The sensors are produced with three layers of *Rogers 4003* substrate as sketched in Figure 5.11. Only the version with the guard electrode is shown. The middle layer contains a milled channel for inserting the MUT. The other two layers contain the electrodes and a ground plane, respectively.

For ease of handling, a 1 mm x 0.5 mm FR4 slab is used as MUT. The permittivity of FR4 is much lower than that of fluids such as water, which decreases the capacitance. The electrodes are designed to ensure impedances that are much higher than

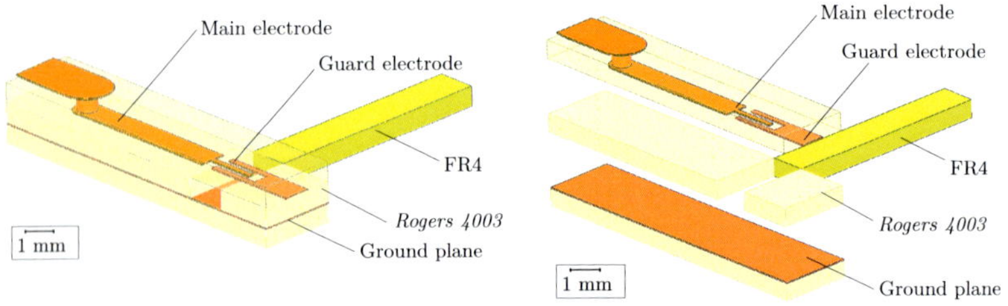

(a) Sensor design. (b) Exploded view.

Figure 5.11: Sketch of the PCB sensor with a guard electrode, implemented with three substrates.

the characteristic impedance of 50 Ω. This is a requirement as described in Section 5.1. Typical impedances measured here are in the order of 10 kΩ.

In order to have different contributions of fringing field, the FR4 slab has two positions, a full and a partial insertion. Both sensors are measured in both positions for comparison. The calibration materials are air and a *Rogers 4003* slab fully inserted into the sensors. This is a similar scenario to the calibration of a microchip sensor with air and a single liquid completely filling the microfluidic channel, respectively.

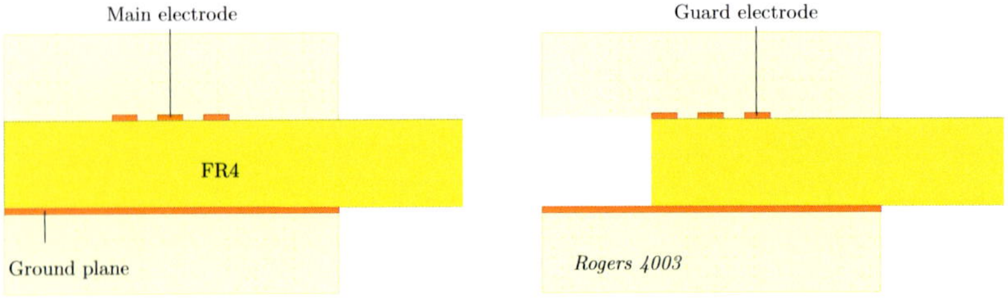

(a) Fully inserted FR4 slab. (b) Partially inserted FR4 slab.

Figure 5.12: Cross-section of the manufactured PCB sensor with two different slab positions.

The different slab positions are sketched in Figure 5.12 and are the same for both sensors. Figure 5.13 shows one of the produced sensors with PCB connectors. The see-through holes enable accurate positioning of the partially inserted slab. The goal is to ensure that some of the fringing field lines go through the air instead of the FR4

(a) Fully inserted FR4 slab. (b) Partially inserted FR4 slab.

Figure 5.13: The manufactured sensor with two different slab positions.

in case of the partial insertion. The reference planes, marked with copper triangles, are set before the electrodes with a TRL calibration.

The measurement results are depicted in Figure 5.14 for both cases. The extracted permittivity of the FR4 without a guard electrode is inaccurate if the slab is partially inserted as seen in Figure 5.14(a). Compared to the 4.4 reference, it is underestimated with values around 3.8. Different positions yield different inaccuracies. With a guard electrode, as shown in Figure 5.14(b), the permittivity is correctly extracted as approximately 4.4 for both fully and partially inserted slabs. With these results, the decoupling of the fringing field is successfully demonstrated in the scaled experiment.

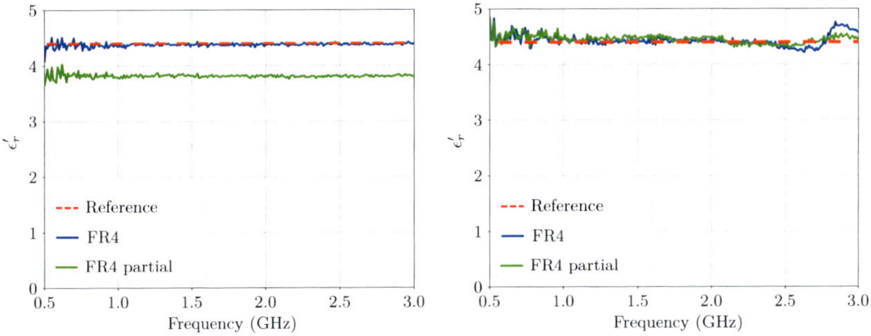

(a) Without a guard electrode. (b) With a guard electrode.

Figure 5.14: Permittivity extracted from the measured S-parameters.

5.4 Microchip Implementation

Figure 5.15 depicts an exemplary cross-section of a microchip for the guard electrode implementation. It is a modified version of the cross-section from Chapter 4. Here, an additional gold layer is available and is placed at the bottom of the top glass substrate. Due to the etching process of the top glass substrate, the available resolution of the top gold layer is significantly lower than for the bottom one, i.e., two electrodes would have to be at least 10 µm apart. For this reason, the electrodes are placed on the bottom glass substrate and the ground plane on the top. Due to the lack of classical via structures as in PCB technology, small holes are made in the top glass substrate and filled with conductive glue. Here, RS Pro silver loaded epoxy conductive adhesive is used. This way, a connection between the metal layers is established, i.e., between the outer contacts of the CPW and the top ground layer.

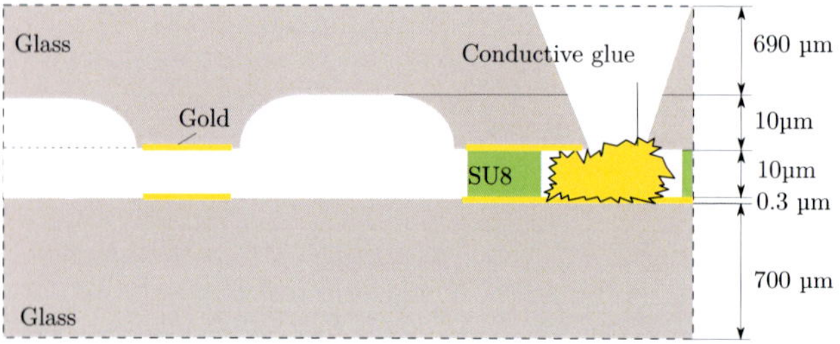

Figure 5.15: Cross-section of the different layers available for the microchip implementation.

Figure 5.16 depicts the sensor design with different layers as drawn in Ansys HFSS. A CPW line is again used as the basis structure. It has the same dimensions as in Chapter 4. The two electrodes are formed with the center conductor line of the CPW and are depicted in Figure 5.16(a). The SU8 part of the channel is shown in Figure 5.16(b). The microfluidic trapping mechanism is the same as in Chapter 4. Figure 5.16(c) contains the ground plane, which is connected with the ground contacts of the CPW. Figure 5.16(d) shows the design flipped upside down, i.e., the top glass substrate is below the metal layers.

Due to manufacturing uncertainties with the novel design, two versions of the electrodes are designed and manufactured. They are depicted in Figure 5.17. One version has wider electrodes, a larger trapping space, and a higher chance of successful man-

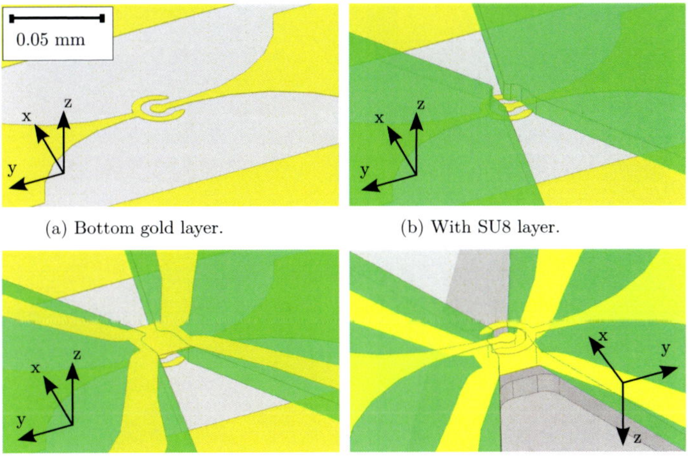

(a) Bottom gold layer. (b) With SU8 layer.

(c) With top gold layer. (d) Flipped, with top glass substrate.

Figure 5.16: Guard electrode sensor design.

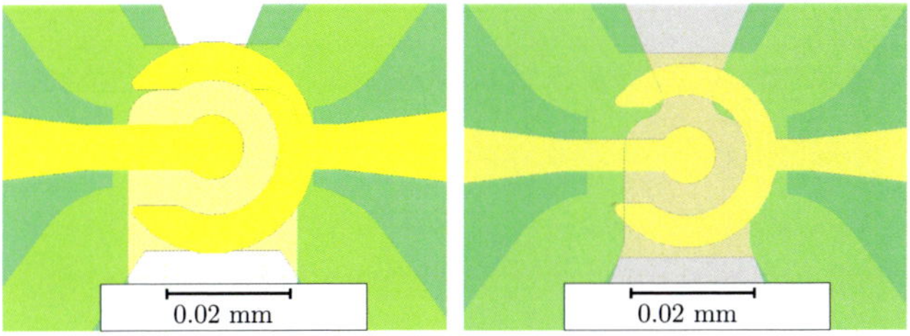

(a) Larger version. (b) Smaller version.

Figure 5.17: Two design versions for manufacturing.

ufacturing. The other one is smaller and pushes the production limits. Indeed, after production on a wafer, only two out of eight microchips with the smaller sensor are usable, whereas four out of eight are usable for the larger version. This make a total of six potentially usable microchips for measurements. Apart from the electrode surface area, the microchips are identical.

One of the six manufactured microchips is shown in Figure 5.18. Figure 5.18(a) shows the whole chip with the two ports and three holes for the microfluidic channel. The

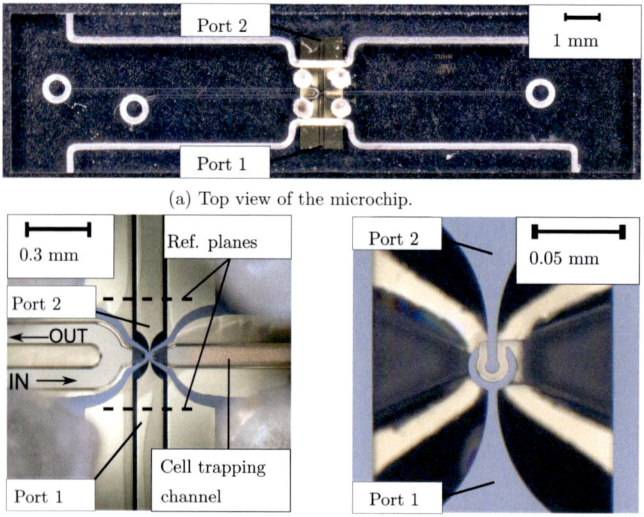

(a) Top view of the microchip.

(b) Top view.

(c) Bottom view. (chip flipped from left to right)

Figure 5.18: Different views on one of the manufactured microchips, 20 mm x 5 mm in size. Between the gold layer and the glass substrate there is typically an adhesion layer of several nm thickness with a different metal, e.g., titanium. For this reason, one side of the gold layer appears to be blue.

four holes in the middle are for the conductive glue. Figure 5.18(b) shows the reference planes, the ground electrode, and the three way microfluidic channel. Figure 5.18(c) contains a close up on the electrodes when the microchip is flipped.

Two of the six microchips were not usable due to failed connection between the metal layers. Figure 5.19 contains a case of a successful and an unsuccessful application of the glue. In Figure 5.19(a), the glue is spread almost evenly, whereas it did not quite reach the metal layers in Figure 5.19(b). In successful cases, the DC resistance between the left and right CPW ground plane is in the order of 4 Ω, while for unsuccessful cases it remains an open circuit.

The measurement setup is shown in Figure 5.20. In Figure 5.20(a), the chip holder, the attached tubing, and probes can be seen. For fluid insertion, syringes (not shown here) are attached to the tubing. Figure 5.20(b) shows a top view of the chip with positioned probe tips on the CPW.

(a)

(b)

Figure 5.19: Examples of a (a) successful and (b) unsuccessful application of the conductive glue.

(a) Chip holder, attached tubing, and measurement probes.

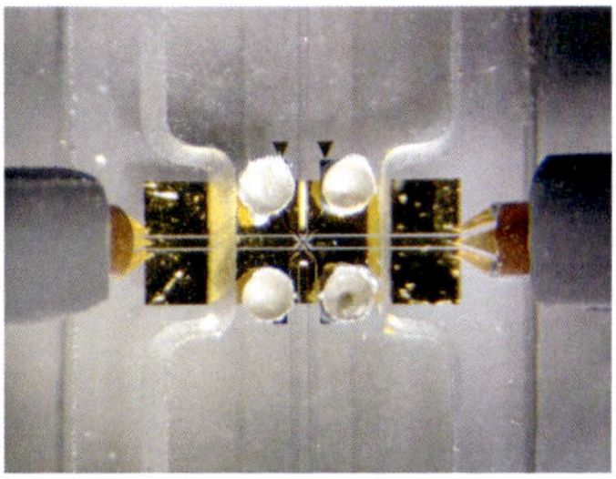

(b) Top view of the chip with measurement probes.

Figure 5.20: Measurement setup.

5.5 Measurements With the Larger Sensor Version

First, solely fluid measurements are done in order to verify the functionality of the novel sensor design. Figure 5.21 depicts the measurement results for three fluids, namely pure ethanol, a 0.9% NaCl, and a 0.5% NaCl solution. Again, the reference values from literature [27], [30] are represented with dotted lines. For the real part of the permittivity in Figure 5.21(a), the measured values are in good agreement with literature. The two NaCl solutions can hardly be distinguished from each other because the real part of their permittivities differ only by about 1. This is lower than the measured variation due to noise. In case of the imaginary part of the permittivity in Figure 5.21(b), all fluids can clearly be distinguished from each other and the measurements are in agreement with literature values.

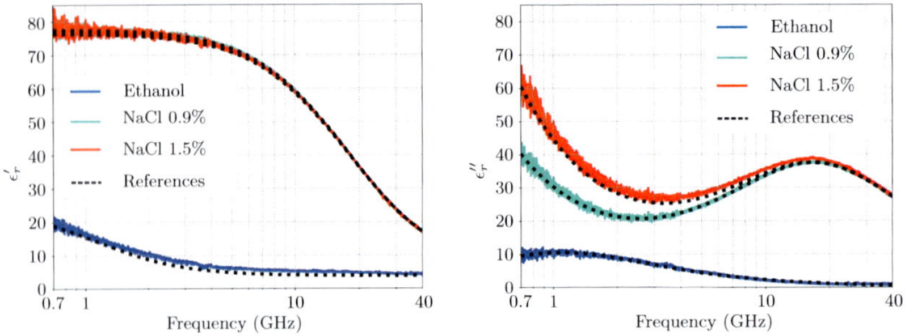

(a) Real part. (b) Imaginary part.

Figure 5.21: Measured permittivity of fluids for sensor validation.

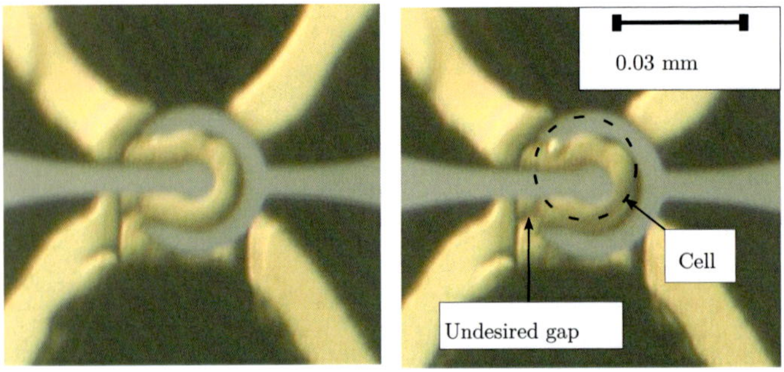

a) Empty sensing area. b) Trapped cell.

Figure 5.22: Trapped cell in the guard electrode sensor.

The cells are prepared in the same way as in Chapter 4. This time, the available living cells for measurements are kept in a culture medium instead of PBS. Figure 5.22 shows a close up look on the sensing area without and with a trapped cell. In the latter case, the cell does not completely fill out the volume between the electrodes, so that some measurement uncertainty due to the undesired gap remains.

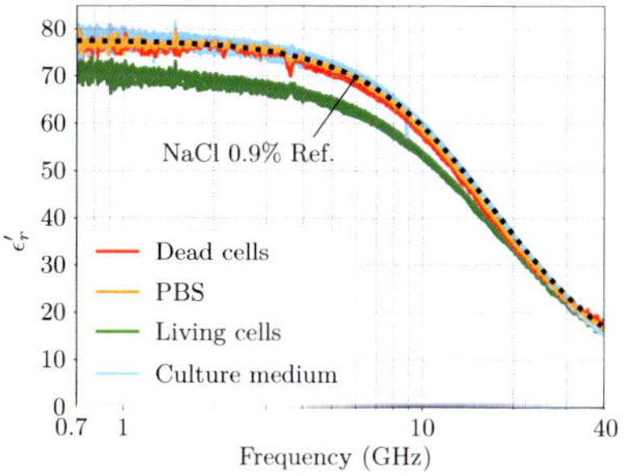

(a) Real part of the permittivity.

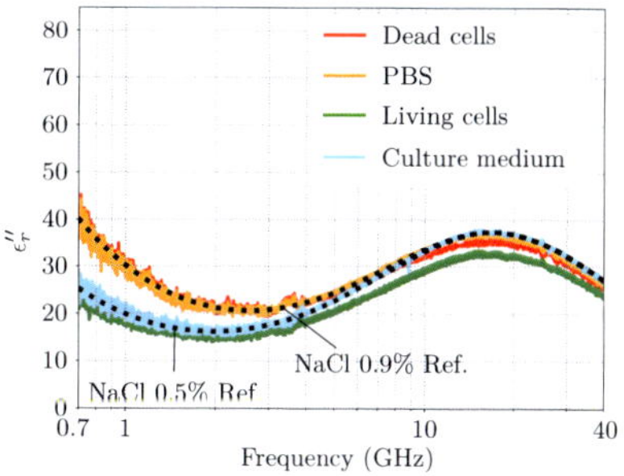

(b) Imaginary part of the permittivity.

Figure 5.23: Measurement results with the guard electrode sensor for living and dead CHO cells, alongside with the measurements of culture medium and PBS.

For the measurement results, three living and two dead CHO cells are measured eleven and four times in total, respectively. The culture medium is measured six times and the PBS thirteen times. All of the curves are plotted in Figure 5.23. Whereas the PBS permittivity is similar to that of a 0.9% NaCl solution, that of the culture medium, here, is similar to that of a 0.5% NaCl solution.

In case of the real part of the permittivity, depicted in Figure 5.23(a), solely the results of the living cells stick out. Those values are the lowest, at around 70 at 1 GHz, while all of the other ones overlap. The imaginary part in Figure 5.23(b) offers some contrast between living and dead cells, compared to their respective medium, in the 10 GHz to 20 GHz range.

The contrasts with the resulting standard deviation between the cells and the mediums are separately plotted in Figure 5.24. For the living cells in Figure 5.24(a), the real part of the permittivity exhibits contrasts of up to 9 below 10 GHz. The imaginary part of the permittivity shows contrasts of up to about 5 above 10 GHz. This is a significantly lower contrast compared to CHO cells measured in Chapter 4, even when taking into account the slightly different surrounding medium (PBS) there. There is a contrast present for dead cells in Figure 5.24(b), but it is not as clear as for living

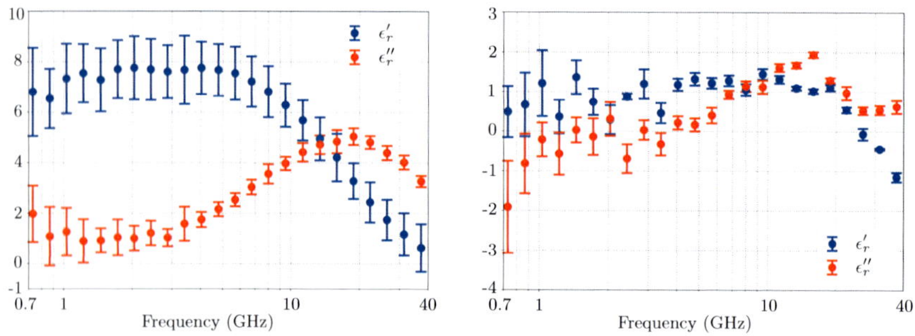

(a) (b)

Figure 5.24: Average contrast and associated standard deviation between (a) living and (b) dead CHO cells, respectively, and the surrounding medium.

cells. For example, at around 15 GHz the contrast has a value of only 2. This is also lower than measured in Chapter 4.

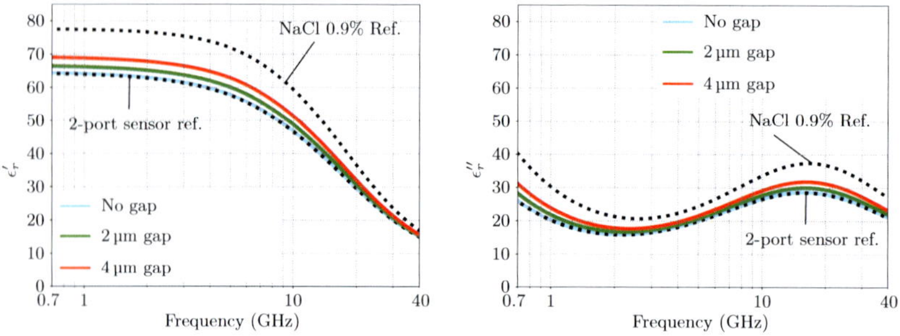

(a) Real part of the permittivity. (b) Imaginary part of the permittivity.

Figure 5.25: Simulated results with different gaps in the sensing area. The material definition of the cell in simulation corresponds to living CHO measurements with the 2-port sensor.

To evaluate the effect of the gap between the captured cell and the microfluidic wall, several simulations are carried out. The material definition for a cell is taken from the 2-port sensor results of a living CHO cell from Chapter 4. The results for three cases, namely with no gap, a 2 µm, and a 4 µm gap, are depicted in Figure 5.25. It can be seen that, for an increasing gap, the permittivity curves move towards the curves of the surrounding medium, which is 0.9% NaCl here. The exact effect on the measured cells can hardly be simulated as the exact shape of the cells in the sensing area is not known. But the tendency is clear.

The measured cells can clearly be distinguished from their medium as evaluated by a t-test in Table 5.1. At 15 GHz, all of the p-values are significantly lower that the 0.01 threshold. Nevertheless, the measured permittivity is not accurate due to the undesired gap.

Table 5.1: p-values for single cell measurements.

Measurements	Frequency	p-value for ϵ'	p-value for ϵ''
Living cells	1 GHz	$6.8 \cdot 10^{-15}$	$4 \cdot 10^{-3}$
	15 GHz	$1.3 \cdot 10^{-13}$	$6 \cdot 10^{-23}$
Dead cells	1 GHz	0.11	0.45
	15 GHz	$2.1 \cdot 10^{-8}$	$1.8 \cdot 10^{-7}$

5.6 Measurements With the Smaller Sensor Version

CHO measurements with the smaller guard electrode sensor version are depicted in Figure 5.26. The results for three living CHO cells are plotted in green. They are lower than those measured with the larger version in Figure 5.23 and more comparable to previous results measured in Chapter 4.

Table 5.2 contains the Cole-Cole parameter of living CHO cells measured with different sensors. It can be seen, that for the lower GHz range, represented by ϵ_s, the guard electrode sensor measured the lowest values, i.e., the highest contrast to the surrounding medium. The values ϵ_∞ and α also show differences, but are inaccurate due the frequency limits of the measurements. The time constant τ is the highest for the guard electrode sensor. The conductance σ stays below 1 for all living cells, but shows no clear tendency between the sensors.

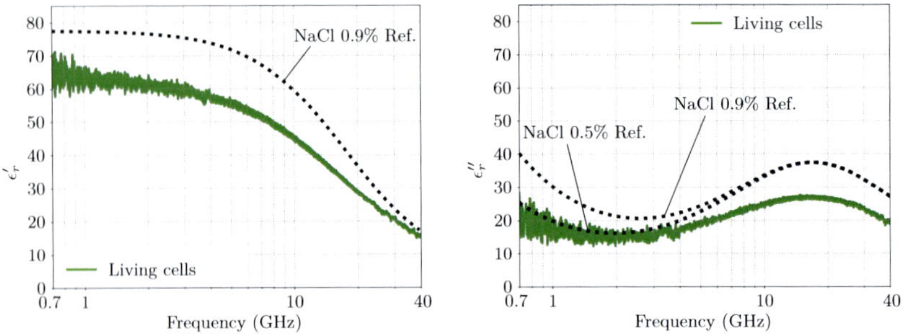

(a) Real part of the permittivity. (b) Imaginary part of the permittivity.

Figure 5.26: Living CHO cells measurements with the smaller guard electrode sensor.

Table 5.2: Cole-Cole model parameters comparison for CHO cells at room temperature, valid between 0.7 GHz and 40 GHz.

	ϵ_s	ϵ_∞	τ	α	σ
1-Port, CHO:					
Living	65.59	4.49	9.35 ps	0.042	0.7 S/m
2-Port, CHO:					
Living	64.6	4.06	9.6 ps	0.059	0.89 S/m
Guard, CHO:					
Living	63.01	4.93	10.03 ps	0.073	0.74 S/m

The results are noisier than for the planar sensors. Nevertheless, when evaluated, e.g., with the Cole-Cole parameter ϵ_s, they indicate a measurable different effective permittivity with the guard electrode sensor compared to those measured with the planar sensors, see Table 5.2. The sensors illuminate different parts of the cell and the measured permittivity difference might be due to different material composition, i.e., at the center of the cell compared to the cell's side. To reduce the noise influence in the future, and to get more accurate results, the measurement bandwidth of the network analyzer should be decreased.

Conclusion

The goal of this thesis is to develop procedures and sensors for broadband single cell permittivity measurements. To achieve this, a single fluid calibration procedure is developed and optimized with various sensors. Different aspects regarding equivalent circuits, parasitic influences, reference plane positions and complex characteristic impedance are analyzed and evaluated. It is demonstrated, that the developed procedure can be used with various technologies, such as OEC, PCB, and microchip technology. Different sensor designs, such as 1-port, 2-port, planar, and multi-layer devices also benefit from the developed calibration procedure. With these insights, microchip sensors for single cell characterization are developed.

Two planar sensors are designed, produced, and evaluated. One is a 1-port device, the other has two ports. The latter type is common in literature with similar designs. There, the presented permittivity information on single cells is limited. Both of these sensors illuminate the cell from the side and utilize a three-way single cell trap. The trapping mechanism is optimized to squeeze the cells vertically and ensure a physical connection to the electrodes. Two different cell lines, namely CHO and HepG2, are measured with these sensors. For both, three cell versions are measured in different pathological states, namely living and dead cells affected by formalin and ethanol, respectively. In all cases, the broadband complex permittivity is successfully determined. Different permittivity footprints are measured for different pathological states. For living cells, the two cell lines can easily be differentiated. The human liver cancer cell line HepG2 shows a higher contrast to the surrounding medium for single cell measurements, compared to the CHO cells.

A novel sensor design with a guard electrode for the illumination of the center of the cell is developed. The guard electrodes concept, which was available for low frequencies, is adapted for microwave measurements. Experimentally, its functionality is successfully demonstrated on an scaled PCB sensor. Two versions of the single cell sensor are designed, produced, and tested. The one with a larger sensing area exhibits some inaccuracy due to an undesired gap between the cell and the microfluidic wall. The smaller version has some production difficulties but is capable of successfully determining the complex permittivity. The permittivity of single living CHO cells is measured and shows higher contrast in the lower GHz range than when measured with the planar sensors. This makes the guard electrode design a promising approach for future developments.

In conclusion, the goal of the thesis has been reached. A suitable calibration procedure is developed and several sensor designs for successful single cell characterization are presented. These results are promising and offer distinctive permittivity footprints for single biological cells of two cell lines. In future, the designs can be optimized to reduce the practical issues like low yield and congestion. Further cell lines can be measured to have even more data for analysis and comparison. Ultimately, clinical studies can be performed on patients suffering from caner. Minimal tissue samples can be cultivated and single cells analyzed like in this work for more medical insights.

List of Abbreviations

CHO Chinese Hamster Ovary

CPW coplanar waveguide

HUVEC human umbilical vein endothelial cells

MUT material under test

OEC open-ended coaxial line

PBS phosphate buffered saline

PCB printed circuit board

TRL transmission-reflection-line

Bibliography

[1] K. Grenier, F. Artis, M. Poupot, J. Foumié, and D. Dubuc, "Label-free discrimination of human lymphoma cell sub-populations with microwave dielectric spectroscopy," in *IEEE MTT-S International Microwave Symposium (IMS)*, Jun. 2018, pp. 907–910.

[2] N. Meyne, G. Fuge, A. P. Zeng, and A. F. Jacob, "Resonant microwave sensors for picoliter liquid characterization and nondestructive detection of single biological cells," *IEEE J. Electromagn., RF, Microw. Med. Biol.*, vol. 1, no. 2, pp. 98–104, Dec. 2017.

[3] X. Ma, X. Du, H. Li, X. Cheng, and J. C. M. Hwang, "Ultra-wideband impedance spectroscopy of a live biological cell," *IEEE Trans. Microw. Theory Techn.*, vol. 66, no. 8, pp. 3690–3696, Aug. 2018.

[4] A. Tamra, D. Dubuc, M. P. Rols, and K. Grenier, "Microwave monitoring of single cell monocytes subjected to electroporation," *IEEE Trans. Microw. Theory Techn.*, vol. 65, no. 9, pp. 3512–3518, Sep. 2017.

[5] C. Arena, M. Sano, J. Rossmeisl Jr, J. Caldwell, P. Garcia, M. Rylander, and R. Davalos, "High-frequency irreversible electroporation (H-FIRE) for non-thermal ablation without muscle contraction," in *Biomedical engineering online*, vol. 10, Nov. 2011, p. 102.

[6] T. de Baere and F. Deschamps, "New tumor ablation techniques for cancer treatment (microwave, electroporation)," *Diagnostic and Interventional Imaging*, vol. 95, no. 7, pp. 677 –682, 2014, Interventional radiology in oncology.

[7] U Kaatze, "Measuring the dielectric properties of materials. Ninety-year development from low-frequency techniques to broadband spectroscopy and high-frequency imaging," *Measurement Science and Technology*, vol. 24, no. 1, 2013.

[8] A. P O'Rourke, M. Lazebnik, J. Bertram, M. Converse, S. C Hagness, J. Webster, and D. M Mahvi, "Dielectric properties of human normal, malignant and cirrhotic liver tissue: In vivo and ex vivo measurements from 0.5 to 20 GHz using a precision open-ended coaxial probe," in *Physics in medicine and biology*, vol. 52, Sep. 2007, pp. 4707–4719.

[9] V. Nerguizian, A. Alazzam, I. Stiharu, and M. Burnier, "Characterization of several cancer cell lines at microwave frequencies," *Measurement*, vol. 109, pp. 354 –358, 2017.

[10] L. Y. Zhang, C. B. M. du Puch, A. Lacroix, C. Dalmay, A. Pothier, C. Lautrette, S. Battu, F. Lalloué, M. O. Jauberteau, and P. Blondy, "Microwave biosensors for identifying cancer cell aggressiveness grade," in *IEEE MTT-S International Microwave Symposium (IMS)*, Jun. 2012, pp. 1–3.

[11] F. Artis, T. Chen, T. Chretiennot, J. J. Fournie, M. Poupot, D. Dubuc, and K. Grenier, "Microwaving biological cells: Intracellular analysis with microwave dielectric spectroscopy," *IEEE Microwave Magazine*, vol. 16, no. 4, pp. 87–96, May 2015.

[12] W. Chen, D. Dubuc, and K. Grenier, "Microwave dielectric spectroscopy of a single biological cell with improved sensitivity up to 40 GHz," in *IEEE MTT-S International Microwave Symposium (IMS)*, May 2015, pp. 1–3.

[13] X. Ma, X. Du, C. R. Multari, Y. Ning, X. Luo, V. Gholizadeh, C. Palego, X. Cheng, and J. C. M. Hwang, "Reproducible broadband measurement for cytoplasm capacitance of a biological cell," in *IEEE MTT-S International Microwave Symposium (IMS)*, May 2016, pp. 1–4.

[14] P. B. Ishai, M. S. Talary, A. Caduff, E. Levy, and Y. Feldman, "Electrode polarization in dielectric measurements: a review," *Measurement Science and Technology*, vol. 24, no. 10, 2013.

[15] J. Ferlay, I. Soerjomataram, R. Dikshit, S. Eser, C. Mathers, M. Rebelo, D. M. Parkin, D. Forman, and F. Bray, "Cancer incidence and mortality worldwide: Sources, methods and major patterns in GLOBOCAN 2012," *International Journal of Cancer:136*, pp. 359–386, 2014.

[16] R. Kurman and I.-M. Shih, "The origin and pathogenesis of epithelial ovarian cancer: A proposed unifying theory," in *The American journal of surgical pathology*, vol. 34, Feb. 2010, pp. 433–43.

[17] J. G. Elmore, M. B. Barton, V. M. Moceri, S. Polk, P. J. Arena, and S. W. Fletcher, "Ten-year risk of false positive screening mammograms and clinical breast examinations," *New England Journal of Medicine*, vol. 338, no. 16, pp. 1089–1096, 1998.

[18] E. Read, J. Park, R. Shah, B. Riley, K. Brorson, and A. Rathore, "Process analytical technology (PAT) for biopharmaceutical products: Part I. concepts and applications," *Biotechnology and Bioengineering*, vol. 105, no. 2, pp. 276–284, 2010.

[19] K P. Jayapal, K. Wlaschin, W. Hu, and M. Yap, "Recombinant protein therapeutics from CHO cells - 20 years and counting," in *Chemical Engineering Progress*, vol. 103, Oct. 2007, pp. 40–47.

[20] K. S. Cole and R. H. Cole, "Dispersion and absorption in dielectrics I. Alternating current characteristics," *The Journal of Chemical Physics*, vol. 9, no. 4, pp. 341–351, 1941.

[21] S. Grimnes and Ørjan G. Martinsen, *Bioimpedance and bioelectricity basics (third edition)*. Academic Press, 2015.

[22] A. Vander Vorst, A. Rosen, and Y. Kotsuka, *RF / Microwave interaction with biological tissues*, ser. Wiley Series in Microwave and Optical Engineering. Wiley, 2006.

[23] G. H. Markx and C. L. Davey, "The dielectric properties of biological cells at radiofrequencies: applications in biotechnology," *Enzyme and microbial technology*, vol. 25, no. 3, pp. 161 –171, 1999.

[24] S. Grimnes and O. Martinsen, "Alpha-dispersion in human tissue," *Journal of Physics: Conference Series*, vol. 224, p. 012 073, May 2010.

[25] A. Savić, *Microwave sensors for the characterization of biological cells*. Master's Thesis, Hamburg University of Technology, Hamburg, Germany, Aug. 2017.

[26] K. V. S. Kaler and T. B. Jones, "Dielectrophoretic spectra of single cells determined by feedback-controlled levitation," *Biophysical Journal*, vol. 57, pp. 173–182, 1990.

[27] C. G. A. Peyman and E. Grant, "Complex permittivity of sodium chloride solutions at microwave frequencies," *Bioelectromagnetics*, vol. 28, pp. 264–274, 2007.

[28] A. Hartman, J. Ekman, G. Antonini, and D. Romano, "A descriptor form implementation of PEEC models incorporating dispersive and lossy dielectrics," in *2016 IEEE International Symposium on Electromagnetic Compatibility (EMC)*, 2016, pp. 206–211.

[29] A. Stogryn, "Equations for calculating the dielectric constant of saline water (correspondence)," *IEEE Trans. Microw. Theory Techn.*, vol. 19, no. 8, pp. 733–736, Aug. 1971.

[30] P. Petong, R. Pottel, and U. Kaatze, "Water-Ethanol mixtures at different compositions and temperatures. A dieletric relaxation study," *The Journal of Physical Chemistry A*, vol. 104, no. 32, pp. 7420–7428, 2000.

[31] G. Brodie, M. Jacob, and P. Farrell, *Microwave and radio-frequency technologies in agriculture: An introduction for agriculturalists and engineers*. De Gruyter Open Poland, Dec. 2015.

[32] A. P. Gregory and R. N. Clarke, "A review of RF and microwave techniques for dielectric measurements on polar liquids," *IEEE Trans. Dielectr. Electr. Insul.*, vol. 13, no. 4, pp. 727–743, Aug. 2006.

[33] J. Sheen, "Comparisons of microwave dielectric property measurements by transmission/reflection techniques and resonance techniques," *Measurement Science and Technology*, vol. 20, no. 4, p. 042 001, 2009.

[34] T. Tosaka, K. Fujii, K. Fukunaga, and A. Kasamatsu, "Development of complex relative permittivity measurement system based on free-space in 220-330-GHz range," *IEEE Trans. THz Sci. Technol.*, vol. 5, no. 1, pp. 102–109, Jan. 2015.

[35] D. Shi, C. Wang, and Y. Gao, "A new permittivity measurement method for walls in indoor scenes," *IEEE Trans. Antennas Propag.*, vol. 67, no. 4, pp. 2118–2129, Apr. 2019.

[36] A. Kraszewski, M. A. Stuchly, and S. S. Stuchly, "ANA calibration method for measurements of dielectric properties," *IEEE Trans. Instrum. Meas.*, vol. 32, no. 2, pp. 385–387, Jun. 1983.

[37] Y. Wei and S. Sridhar, "Technique for measuring the frequency-dependent complex dielectric constants of liquids up to 20 GHz," *Review of Scientific Instruments*, vol. 60, no. 9, pp. 3041–3046, 1989.

[38] Y. Z. Wei and S. Sridhar, "Radiation-corrected open-ended coax line technique for dielectric measurements of liquids up to 20 GHz," *IEEE Trans. Microw. Theory Techn.*, vol. 39, no. 3, pp. 526–531, Mar. 1991.

[39] S. Julrat and S. Trabelsi, "Open-ended coplanar waveguide sensor for dielectric permittivity measurement," in *IEEE International Instrumentation and Measurement Technology Conference (I2MTC)*, May 2018, pp. 1–5.

[40] M. D. Janezic and J. A. Jargon, "Complex permittivity determination from propagation constant measurements," *IEEE Microwave and Guided Wave Letters*, vol. 9, no. 2, pp. 76–78, Feb. 1999.

[41] J. C. Booth, N. D. Orloff, J. Mateu, M. Janezic, M. Rinehart, and J. A. Beall, "Quantitative permittivity measurements of nanoliter liquid volumes in microfluidic channels to 40 GHz," *IEEE Trans. Instrum. Meas.*, vol. 59, no. 12, pp. 3279–3288, Dec. 2010.

[42] V. Parshin, M. Tretyakov, M. Koshelev, and E. Serov, "Modern resonator spectroscopy at SubMM wavelengths," in *2013 International Kharkov Symposium on Physics and Engineering of Microwaves, Millimeter and Submillimeter Waves*, Jun. 2013, pp. 73–75.

[43] N. Meyne, S. Latus, and A. F. Jacob, "Corrugated coplanar transmission-line sensor for broadband liquid sample characterization," in *German Microwave Conference (GeMiC)*, Mar. 2014, pp. 1–4.

[44] N. Meyne, W. Müller-Wichards, H. K. Trieu, and A. F. Jacob, "Quasi-lumped coplanar transmission-line sensors for broadband liquid characterization," in *European Microwave Conference (EuMC)*, Oct. 2014, pp. 687–690.

[45] N. Meyne née Haase, G. Fuge, H. K. Trieu, A. P. Zeng, and A. F. Jacob, "Miniaturized transmission-line sensor for broadband dielectric characterization of biological liquids and cell suspensions," *IEEE Trans. Microw. Theory Techn.*, vol. 63, no. 10, pp. 3026–3033, Oct. 2015.

[46] E. Tanabe and W. T. Joines, "A nondestructive method for measuring the complex permittivity of dielectric materials at microwave frequencies using an open transmission line resonator," *IEEE Trans. Instrum. Meas.*, vol. IM-25, no. 3, pp. 222–226, Sep. 1976.

[47] T. Chretiennot, D. Dubuc, and K. Grenier, "A microwave and microfluidic planar resonator for efficient and accurate complex permittivity characterization of aqueous solutions," *IEEE Trans. Microw. Theory Techn.*, vol. 61, no. 2, pp. 972–978, Feb. 2013.

[48] N. Haase and A. F. Jacob, "Resonant substrate-integrated near-field sensors with improved sensitivity," in *European Microwave Integrated Circuit Conference*, Oct. 2013, pp. 468–471.

[49] ——, "Substrate-integrated half-mode resonant near-field sensor for liquid characterization," in *European Microwave Conference (EuMC)*, Oct. 2013, pp. 310–313.

[50] A. A. Helmy and K. Entesari, "A 1–8-GHz miniaturized spectroscopy system for permittivity detection and mixture characterization of organic chemicals," *IEEE Trans. Microw. Theory Techn.*, vol. 60, no. 12, pp. 4157–4170, 2012.

[51] X. Bao, I. Ocket, J. Bao, D. Kil, L. Brancato, and B. Nauwelaers, "Broadband dielectric spectroscopy measurements of liquids combining interdigital capacitor and coplanar waveguide," in *European Microwave Conference (EuMC)*, Oct. 2017, pp. 946–949.

[52] K. Grenier, D. Dubuc, P. Poleni, M. Kumemura, H. Toshiyoshi, T. Fujii, and H. Fujita, "Integrated broadband microwave and microfluidic sensor dedicated to bioengineering," *IEEE Trans. Microw. Theory Techn.*, vol. 57, no. 12, pp. 3246–3253, Dec. 2009.

[53] F. Artis, D. Dubuc, J. J. Fournié, M. Poupot, and K. Grenier, "Microwave dielectric spectroscopy of cell membrane permeabilization with saponin on human B lymphoma cells," in *IEEE MTT-S International Microwave Symposium (IMS)*, Jun. 2014, pp. 1–4.

[54] N. Meyne, G. Fuge, S. Hemanth, H. K. Trieu, A. P. Zeng, and A. F. Jacob, "Broadband dielectric characterization of CHO-K1 cells using miniaturized transmission-line sensor," in *IEEE Topical Conference on Biomedical Wireless Technologies, Networks, and Sensing Systems (BioWireleSS)*, Jan. 2015, pp. 1–3.

[55] S. Liu, N. D. Orloff, C. A. E. Little, W. Zhao, J. C. Booth, D. F. Williams, I. Ocket, D. M. M. P. Schreurs, and B. Nauwelaers, "Hybrid characterization of nanolitre dielectric fluids in a single microfluidic channel up to 110 GHz," *IEEE Trans. Microw. Theory Techn.*, vol. 65, no. 12, pp. 5063–5073, Dec. 2017.

[56] N. Haase and A. F. Jacob, "Characterization of biological substances using a substrate integrated microwave near-field sensor," in *European Microwave Conference (EuMC)*, Oct. 2012, pp. 432–435.

[57] N. Haase and A. F. Jacob, "Dielectric contrast measurements on biological substances with resonant microwave near-field sensors," *International Journal of Microwave and Wireless Technologies*, vol. 5, no. 3, 221–230, 2013.

[58] N. Meyne, *Microwave sensors for non-destructive characterization of liquids and biological cells.* Dissertation, Hamburg University of Technology, Hamburg, Germany, 2020.

[59] T. Chen, D. Dubuc, M. Poupot, J. J. Fournie, and K. Grenier, "Accurate nanoliter liquid characterization up to 40 GHz for biomedical applications: Toward noninvasive living cells monitoring," *IEEE Trans. Microw. Theory Techn.*, vol. 60, no. 12, pp. 4171–4177, Dec. 2012.

[60] T. Chen, F. Artis, D. Dubuc, J.-J. Fournié, M. Poupot, and K. Grenier, "Microwave biosensor dedicated to the dielectric spectroscopy of a single alive biological cell in its culture medium," in *IEEE MTT-S International Microwave Symposium (IMS)*, 2013, pp. 1–4.

[61] A. Zedek, D. Dubuc, and K. Grenier, "Microwave permittivity extraction of individual biological cells submitted to different stimuli," in *IEEE MTT-S International Microwave Symposium (IMS)*, Jun. 2017, pp. 865–868.

[62] A. Tamra, M. Deburghgraeve, D. Dubuc, M. P. Rols, and K. Grenier, "Microwave dielectric spectroscopy for single cell irreversible electroporation monitoring," in *IEEE MTT-S International Microwave Symposium (IMS)*, May 2016, pp. 1–4.

[63] X. Du, X. Ma, L. Li, H. Li, X. Cheng, and J. C. M. Hwang, "Ultra-wideband characterization, electroporation, and dielectrophoresis of a live biological cell using the same vector network analyzer," in *IEEE MTT-S International Microwave Symposium (IMS)*, Jun. 2018, pp. 1148–1151.

[64] X. Du, C. Ladegard, X. Ma, X. Cheng, and J. C. M. Hwang, "Ultra-wideband electrical sensing of nucleus size in a live cell," in *European Microwave Conference (EuMC)*, 2019, pp. 208–211.

[65] H. Morgan, T. Sun, D. Holmes, S. Gawad, and N. G. Green, "Single cell dielectric spectroscopy," *Journal of Physics D: Applied Physics*, vol. 40, no. 1, p. 61, 2007.

[66] Y. Cho, H. S. Kim, A. B. Frazier, Z. G. Chen, D. M. Shin, and A. Han, "Whole-cell impedance analysis for highly and poorly metastatic cancer cells," *Journal of Microelectromechanical Systems*, vol. 18, no. 4, pp. 808–817, Aug. 2009.

[67] Y. Zheng, E. Shojaei-Baghini, C. Wang, and Y. Sun, "Rapid measurement of specific membrane capacitance and cytoplasm conductivity on single cells," in *IEEE International Conference on Micro Electro Mechanical Systems (MEMS)*, Jan. 2013, pp. 1105–1108.

[68] A. Savić, N. Meyne, and A. F. Jacob, "Microwave impedance sensors for the dielectric characterization of liquids," in *German Microwave Conference (GeMiC)*, Mar. 2018, pp. 219–222.

[69] A. Savić, A. Stepputtis, and A. F. Jacob, "Impedance sensor for fluid permittivity measurements up to 20 GHz with single-liquid calibration," in *European Microwave Conference (EuMC)*, Sep. 2018, pp. 227–230.

[70] A. Savić, N. Meyne, and A. F. Jacob, "Model-based microwave dielectroscopy of fluids with impedance sensors," *IEEE Trans. Microw. Theory Techn.*, pp. 1–9, 2019.

[71] A. Nyshadham, C. L. Sibbald, and S. S. Stuchly, "Permittivity measurements using open-ended sensors and reference liquid calibration – An uncertainty analysis," *IEEE Trans. Microw. Theory Techn.*, vol. 40, no. 2, pp. 305–314, Feb. 1992.

[72] C. D. Abeyrathne and E. Skafidas, "Complex permittivity measurements in 1–30 GHz using a MEMS probe," *Journal of Microelectromechanical Systems*, vol. 24, no. 4, pp. 976–981, Aug. 2015.

[73] S. Schmidt, M. Schüßler, and R. Jakoby, "All liquid based calibration scheme for microwave dielectrometry," in *IEEE MTT-S International Microwave Symposium (IMS)*, Jun. 2017, pp. 1266–1269.

[74] K. Kim, N. Kim, S. H. Hwang, Y. K. Kim, and Y. Kwon, "A miniaturized broadband multi-state reflectometer integrated on a silicon MEMS probe for complex permittivity measurement of biological material," *IEEE Trans. Microw. Theory Techn.*, vol. 61, no. 5, pp. 2205–2214, May 2013.

[75] X. Bao, I. Ocket, D. Kil, J. Bao, R. Puers, and B. Nauwelaers, "Liquid measurements at microliter volumes using 1-port coplanar interdigital capacitor," in *IEEE MTT-S International Microwave Bio Conference (IMBioC)*, May 2017, pp. 1–4.

[76] X. Bao, I. Ocket, G. Crupi, D. Schreurs, J. Bao, D. Kil, B. Puers, and B. Nauwelaers, "A planar one-port microwave microfluidic sensor for microliter liquids characterization," *IEEE J. Electromagn., RF, Microw. Med. Biol.*, vol. PP, no. 99, 2018.

[77] R. B. Marks and D. F. Williams, "Characteristic impedance determination using propagation constant measurement," *IEEE Microwave and Guided Wave Letters*, vol. 1, no. 6, pp. 141–143, Jun. 1991.

[78] D. F. Williams and R. B. Marks, "Transmission line capacitance measurement," *IEEE Microwave and Guided Wave Letters*, vol. 1, no. 9, pp. 243–245, Sep. 1991.

[79] C. Gabriel, T. Y. A. Chan, and E. H. Grant, "Admittance models for open ended coaxial probes and their place in dielectric spectroscopy," *Physics in Medicine & Biology*, vol. 39, no. 12, pp. 2183–2200, 1994.

[80] C. D. Abeyrathne, M. N. Halgamuge, P. M. Farrell, and E. Skafidas, "An ab-initio computational method to determine dielectric properties of biological materials," *Scientific Reports*, vol. 3, p. 1796, 2013.

[81] R. B. Marks, "Multi-Line Calibration for MMIC Measurement," in *36th ARFTG Conference Digest*, vol. 18, 1990, pp. 47–56.

[82] C. Wu, Y. Xu, J. Li, and S. Gao, "Effects of the length of thru on the measurement precision in TRL technique," *IEEE Microw. Wireless Compon. Lett.*, vol. 24, no. 12, pp. 905–907, Dec. 2014.

[83] C. Wu, L. Sima, X. Hou, and J. Li, "Periodical measurement error with respect to the length of thru in TRL technique," *IEEE Trans. Microw. Theory Techn.*, vol. 65, no. 5, pp. 1605–1614, May 2017.

[84] K. Levenberg, "A method for the solution of certain non-linear problems in least squares," *Quarterly of Applied Mathematics*, vol. 2, pp. 164–168, 1944.

[85] D. W. Marquardt, "An Algorithm for Least-Squares Estimation of Nonlinear Parameters," *Journal of the Society for Industrial and Applied Mathematics*, vol. 11, no. 2, pp. 431–441, 1963. (visited on 10/09/2022).

[86] J. Zhu, J. Shang, T. Olsen, K. Liu, D. Brenner, and Q. Lin, "A mechanically tunable microfluidic cell-trapping device," *Sensors and actuators A: Physical*, vol. 215, pp. 197 –203, 2014.

[87] F. Yesilkoy, R. Ueno, B. X. E. Desbiolles, M. Grisi, Y. Sakai, B. J. Kim, and J. Brugger, "Highly efficient and gentle trapping of single cells in large microfluidic arrays for time-lapse experiments," *Biomicrofluidics*, vol. 10, no. 1, 2016.

[88] A. Savić, F. Freiberger, R. Pörtner, and A. F. Jacob, "A capacitive microwave sensor with guard electrode for single cell characterization," *IEEE J. Electromagn., RF, Microw. Med. Biol.*, 2021.

[89] S. Greenland, S. Senn, K. Rothman, J. Carlin, C. Poole, S. Goodman, and D. Altman, "Statistical tests, P values, confidence intervals, and power: A guide to misinterpretations," *European Journal of Epidemiology*, vol. 31, Apr. 2016.

[90] E. Piuzzi, S. Chicarella, A. Cataldo, E. D. Benedetto, and G. Cannazza, "Design, realization, and experimental characterization of an admittance cell for low-frequency dielectric permittivity measurements on liquids," *IEEE Trans. Instrum. Meas.*, vol. 65, no. 1, pp. 104–111, Jan. 2016.

[91] M. Ahmadi, R. Rajamani, and S. Sezen, "Transparent flexible active Faraday cage enables in vivo capacitance measurement in assembled microsensor," *IEEE Sensors Letters*, vol. 1, no. 5, pp. 1–4, Oct. 2017.

[92] W. C. Heerens and F. C. Vermeulen, "Capacitance of Kelvin guard-ring capacitors with modified edge geometry," *Journal of Applied Physics*, vol. 46, no. 6, pp. 2486–2490, 1975.

[93] A. Savić and A. F. Jacob, "A capacitive microwave sensor with guard electrodes for biological cell characterization," in *IEEE MTT-S International Microwave Biomedical Conference (IMBioC)*, 2020, pp. 1–3.

Curriculum Vitae

Name Savić

First Name Aleksandar

Birth date 15.11.1991

Birth place Banja Luka, Bosnia and Herzegovina (BiH)

09/1998–06/2006	**E**lementary school Mladen Stojanović, Laktaši, BiH
09/2006–06/2007	**E**lectrotechnical school Nikola Tesla, Banja Luka, BiH
10/2007–07/2009	**T**heodor-Heuss-Realschule, Preetz, Germany
09/2009–06/2012	**B**erufliches Gymnasium, Preetz, Germany
10/2012–07/2015	**S**tudies at Hamburg University of Technology, Hamburg Degree: Bachelor of Science
07/2015–09/2015	**I**nternship at Philips GmbH Innovative Technologies, Hamburg
10/2015–08/2017	**S**tudies at Hamburg University of Technology, Hamburg Degree: Master of Science
04/2016–03/2017	**W**orking student at Philips GmbH Innovative Technologies Hamburg
09/2017–09/2021	**P**art-time programmer and consultant TuTech Innovation GmbH, Hamburg
09/2017–12/2021	**R**esearch associate at the Institute of High-Frequency Technology Hamburg University of Technology, Hamburg
01/2022–02/2022	**F**inalizing Dissertation
03/2022–today	**E**MC Test Engineer Airbus Operations GmbH, Hamburg